BIRDING AROUND THE YEAR

THE WILEY NATURE EDITIONS

At the Water's Edge: Nature Study in Lakes, Streams, and Ponds
  by Alan M. Cvancara

Mountains: A Natural History and Hiking Guide
  by Margaret Fuller

The Oceans: A Book of Questions and Answers
  by Don Groves

*Birding Around the World: A Guide to Observing Birds
Everywhere You Travel*
  by Aileen Lotz

*Walking the Wetlands: A Hiker's Guide to Common Plants and
Animals of Marshes, Bogs, and Swamps*
  by Janet Lyons and Sandra Jordan

*Nature Nearby: An Outdoor Guide to America's 25 Most
Visited Cities*
  by Bill McMillon

*Wild Plants of America: A Select Guide for the Naturalist
and Traveler*
  by Richard M. Smith

BIRDING AROUND THE YEAR

When to Find Birds in North America

Aileen Lotz

WILEY

Wiley Nature Editions

JOHN WILEY & SONS, INC.

New York • Chichester • Brisbane • Toronto • Singapore

Unless otherwise noted, all photographs have been provided by the author.

Chapter 1: Pack on his back, binoculars at the ready, birder Jack Holmes is on his way
Chapter 2: California's shining seacoast
Chapter 3: Rocky Mountain streams begin to thaw
Chapter 4: Icebergs on the Churchill River signal summer
Chapter 5: Hawk watch at Cape May, New Jersey
Chapter 6: Early snows in Rocky Mountains
Chapter 7: Lowland coastal marsh in South Carolina
Appendix 1: Killdeer affecting broken wing display
Appendix 2: Upended binoculars show details in tiny flowers

Library of Congress Cataloging-in-Publication Data

Lotz, Aileen R.
    Birding around the year : when to find birds in North America /
Aileen R. Lotz.
        p.   cm. — (Wiley nature editions)
    Bibliography : p.
    ISBN 0-471-62076-9. — ISBN 0-471-51049-1 (pbk.)
    1. Bird watching—North America—Guide-books.   I. Title.
II. Series.
QL681.L67   1989
598'.072347—dc20   89-33570
                                                          CIP

Printed in the United States of America

89 90 10 9 8 7 6 5 4 3 2 1

Foreword

*B*irding can be defined in many ways: as a casual hobby, an avocation, a sport, or even a game. For me, what started out as a casual hobby has turned into an intense challenge as I try to see every species of bird this great continent has to offer.

I thrive on the excitement of the chase, and the exhilaration of finding the bird I am looking for. That feeling of elation, almost euphoria, gets better and better now that there are fewer and fewer species still unseen. Birding is a never-ending challenge.

Being successful in any endeavor requires practice, patience, hard work, a measure of luck, and a strategy. Successful birding requires being in the right place at the right time—being where the bird is, binoculars in hand.

A strategy that helps birders find sought-after birds is one to be applauded. I am delighted to applaud Aileen Lotz who has created in this book an enhanced sense of good timing, along with a full measure of good humor. When and where are the best times and places are the questions she emphasizes. Her book is an important contribution to North American birding literature.

The sheer size and diversity of the North American continent provides never-ending interest and challenge for birders. Distances being so great, it is difficult to see every bird that is resident or shows up on our shores. The most recent edition of the American Birding Association checklist contains 858 species. It is doubtful that anyone will ever see all.

Although most birders view me as a lister, my goal is the total enjoyment of the bird, be it at a feeder in Chattanooga, Tennessee, or on the tundra at distant Attu Island, Alaska. Listing adds to the fun, and I stress fun, for this is my form of enjoyment. The total experience of the "quest" is inscribed indelibly in my memory of years of birding. If I miss a species, I'll try again, but along the way, special times are shared with special friends.

Several milestones mark this path of enjoyment. One is often "lead on" by the century milestones. Your first 100, 300, and 500 species are goals that fire the innate drive to always try for the next bird. These stepping stones led me on to 600 and then 700. Now, for me, it is on to 800. Having reached 795 by the end of 1988, 1989 was off to a good start on my way to the Dry Tortugas when I stopped in Miami to see the Bahama Swallow, an infrequent visitor to our shores. That leaves only four exciting new species to go.

You may not aspire to such achievement as you pursue your interest in birding. If you do, my advice is to "run for the rarities." Searching for the rare, more difficult species will result in seeing the more common birds along the way.

Whatever your level of interest in the sport of birding might be, I have no hesitancy in recommending this book by Aileen Lotz. It will enhance your enjoyment of traveling this continent, and of identifying and enjoying its many birds.

Benton Basham
North America's No. 1 Birder

Preface

*B*irding is fun. This book, hopefully, reflects the enjoyment of visiting the near and far reaches of the North American continent and of seeing some of the most beautiful creatures in the world. In my days as an *early birder,* I was satisfied to see the birds that crossed my path at the time and place I happened to be. Gradually it dawned on me that if I wanted to see a particularly enticing bird, I had to cross the bird's path, to be where the bird would be at the time of year the bird would be there. There are countless books describing the appearance and habits of birds and as many providing details of places to find birds. Information on calendar occurrences has been difficult to ferret out.

Organized by season, this book is designed to fill that gap, particularly for early birders, those just getting started in the sport or hobby of birding. Seasonal chapters, the heart of the book, are followed by a chapter covering those delightful areas of the continent that provide especially good birding at any time of the year. To reinforce the importance of timing, months within the seasonal chapters appear in *ITALIC* type in the text to allow rapid skimming for priority times. These chapters provide a general movement through the seasons, from early to late, highlighting the "hot spots" where birding is particularly good at those times of year.

The significance of any month within a season varies from season-to-season, species-to-species, and place-to-place. Generally, a specific month within a season is most important in the spring when birds are moving steadily northward across the face of the continent, concentrating along the major flyways. In summer, birds have arrived at their nesting destinations and they stay put for a while. The fall migration is more diffused than in the spring, birds aren't creating the song splash of springtime but hawks and waterfowl can be viewed in impressive numbers. It is the fall raptor movement that lures birders to official and unofficial hawk watches. By winter, birds have reached their resting "digs" and birding activity is somewhat more leisurely. The Christmas Bird Count, sponsored by the National Audubon Society, provides continent-wide interest, but the focus of winter birding activity tends to be southerly rather than northerly.

Seasonal chapters begin with a brief definitio of the season. In order not to give the reader armchair jet lag by covering July all across the map, some geographic grouping of regions will be apparent. These chapters conclude with the icing on the cake: rare birds that reach our shores. Although early birders are concentrating on learning the field marks of the common birds and may have only a passing interest in the uncommon and rare birds, it is the excitement of seeing these species that quickens the heart and pace of many birders.

A general introduction to the seasonal approach and a brief overview of the birder's universe is covered in Chapter 1. Chapter 2 focuses on the environment of North America relevant to both bird and birder/traveler. Concluding Appendices provide useful tools especially for the early birder. Appendix 1 covers some of the special lingo used by the birding community and some background information on the families of birds to help place the species in the accepted context. Appendix 2 provides useful information on the who, what, and where of this rapidly growing outdoor activity. This list of books and periodicals should help beginner and expert alike who may be in search of more information. For quick reference, key birding hot spots are located on the map of North America on page 221.

No attempt has been made to mention every single species, or to list every place that provides good birding. Well-known birding "hot spots" have been covered and suggestions included for generic places. The book does not, nor could it, provide detailed information on geographic distribution of the Black-billed Magpie; explicit information on which side road to take to reach the old pine woods required by the Red-cockaded Woodpecker; or to list all the likely species to be seen on your vacation trip to the great Northwest. Such detailed information is available for specific areas in other publications referred to in the text and identified in Appendix 2. If this book whets your appetite to read and research further, then it will have achieved one of its objectives.

The book is designed to be entertaining. I hope its *when*-to-find approach will provide a useful complement to the many *where*-to-find guides, and that its informal style will bring you good birding enjoyment throughout the year.

Acknowledgments

*C*reation of a book such as this is the result of hundreds of thoughts and ideas contributed over the years by a host of birding friends and acquaintances. Most often, it is impossible to trace the origin of a line of thought, a bit of instruction, or a word providing insight concerning a bird's habit or habitat. I am grateful for the contributions and encouragement from many birding friends. I should like to express my thanks to a few for the special help and advice they gave me.

To Bill Bouton, who called on his considerable North American birding experience in correcting, expostulating, cajoling, and otherwise making invaluable suggestions on early rough drafts.

To Victor Emanuel, one of my first birding tour leaders, who made excellent suggestions, comments, and corrections on the first completed draft.

To longtime birder friends Tom and Mary Wood, who read through and commented on the manuscript in its early stages.

To Peter Alden whose permission to quote from his extensive notes was graciously given, and who spotted what to him were obvious errors in the final draft.

To birding buddy Maida Maxham, who helped develop the concept for the book, read parts of early drafts, and who keeps me honest in my listing.

To Jack Holmes who read portions of the manuscript and who helped alleviate some of the photographic problems, my special appreciation.

To Mary Kennan, my first editor, who suggested that I write a book, and to my second editor, David Sobel, who agreed.

To other friends who helped significantly by reading and commenting on various portions of the text—Benton Basham, Peter Carlton, Jim Griffith, Robert Kelley, Gene Miller, Bruce Neville, Arnold Small—I also express my thanks.

In spite of the help from knowledgeable friends, the book contains inadvertent mistakes and shortcomings which are my full responsibility. I will welcome any comments and corrections.

Quotations from *American Birds,* vols. 40 (1986), 41 (1987), and 42 (1988), are courtesy of Susan Roney Drennan, ed. *American Birds* is the ornithological journal published by the National Audubon Society.

Contents

1. **Have Binoculars, Will Travel**   1

   Seasons in the Sun   3
   People Who Bird   9
   To List or Not to List   13
   Do It Alone or with a Group   16

2. **From Sea to Shining Sea**   19

   Arctic to Aruba   20
   Zero in on Zoogeography   21
   Flying the Flyways   28
   The Good Neighbor Policy   31
   The World of Birders   33
   Run for the Rarities   37

3. **Spring: A Time to Move North**   43

   When Is Spring?   45
   Early Birding   46
   The Main Show   50
   Sideshows   59
   Run for the Rarities   72

4. **Summer: A Time for Nesting**   75

   When Is Summer?   77
   Northern Lights   77
   Upper Lower 48   91
   Run for the Rarities   99

5. **Fall: A Time to Move South**   101

   When Is Fall?   102
   Hawking "Hot Spots"   104
   Ambling along the Atlantic   112
   Midcountry Flyways   114
   Plying the Pacific   117
   Southern Living   121
   Monarch's Time   123
   Run for the Rarities   123

6. **Winter: A Time for Resting**   125

   When Is Winter?   127
   Christmas Counting   127
   Snowbirding in the Sunbelt   130
   Tropical Delights   137
   Cold "Hot Spots"   139
   Run for the Rarities   142

7. **A Place for All Seasons**   145

   Super-States   146
   Culling California   148
   Tarry in Texas   152
   Around the Year in Arizona   154
   Sunning, Tanning, and Birding in South Florida   156
   East Coasting   159
   Northwestern Larking   162
   Year-Round Sampler   163
   Hailing Hawaii   165

Appendix 1   **A Little Latin and Lingo**   167

   Put the Pigeons into the Right Holes:
      A Brief Latin Lesson   168
   Birder's Lingo: Learn the Language   173

Appendix 2   **Basic Books and Other Aids to
                Better Birding**   179

   Birding Books and Periodicals   180
   Birding Book Dealers   190
   Birder Friendly Accommodations   194
   Tour Operators   200

The Well-Equipped Birder   208
Rare Bird Alerts   214
Photography   215
General Education   217
Common Sense   218
Important Addresses for Birders   220
Map of Selected Birding Hot Spots in
    North America   221

Bibliographic Summary   **223**

Glossary   **225**

Index   **231**

BIRDING AROUND THE YEAR

One

Have Binoculars,
Will Travel

*A*cross the flat prairie, I could see three indistinct shapes against a small grove of live oak trees. Quickly, I pulled off the road, grabbed my binoculars, and got out of the car. Peering back towards the trees I saw clearly what I had surmised—Sandhill Cranes. Motionless, blending in with the gray trees dripping Spanish moss, the tall gray birds stood at attention, alert to my presence. Three red caps atop three small, ostrichlike heads were distinctly visible. Good sighting.

Binoculars and bird guide on the top of the carry-on, luggage packed, I am about to catch a plane to New York City to meet my editor. Binoculars? New York City? Believe it or not, there are birds there, right in Central Park, in the heart of Manhattan. I might "pick up" a migrating Prothonotary Warbler. Later, I'll take a late afternoon plane to Boston and a weekend of birding with a friend at the Parker River National Wildlife Refuge just beyond the delightful town of Newburyport.

Cincinnati, Ohio, in August would not be a birder's choice, but in a postconference walk, the tall oak trees in a park held some passing interest. I counted Eastern Wood Pewees, American Goldfinches, Yellow-throated Warblers, Purple Finches, and a fair number of other eastern species.

Any time of year will do, wherever you are. No matter where you go, no matter why, the purposeful person will find birds to enjoy. Take your binoculars; don't leave home without them. For birders, binoculars are extensions of their eyes. They see the Sandhill Cranes in central Florida while other travelers whiz by a seemingly empty landscape. For the generally curious, those elk over on the Colorado hillside are worth closer examination. For the lost, binoculars help read distant directional signs.

People moving beyond the backyard bird-watching stage, often do the home territory pretty well. If traveling is not their preference or possibility, they may prefer to learn well the birds of hometown, state, or province. They may become experts to whom visiting birders turn for the best places and times of year to see the local specialties. They prefer to watch the seasons change, to clock the arrival or departure of migrating birds, or to welcome back each year the same family of Purple Martins.

This book may not appeal to such happy hometown birders. Rather, it is intended for the watcher of birds who seeks feathers on far horizons. It is designed for those travelers and birders who wish to sample the richness of the North American continent in its history, diverse scenery and habitat, and wonderful wildlife. This book is written primarily for the North American birder, but with so many birders visiting from other countries, it will be useful to them as well.

Birders? Birding? These terms are used throughout this book. You may stoutly insist on being a *birdwatcher,* although a *National Wildlife* issue a few years ago headlined an article, "They're Not Called 'Bird-watchers' Anymore." British and other Europeans prefer to be bird-watchers while North Americans now prefer to identify themselves as birders. Birder or birdwatcher, you are one of 62 million Americans who, at a minimum, put seed in bird feeders, according to Mike Lipske who wrote that article. *Bird'n Hand News,* a publication for feeder-fillers, reported in its Fall 1988 issue that in 1985 American watchers of birds spent $1 billion on bird seed, another $239 million for bird-houses and such accouterments, and $373 million for binoculars. It's a good business to be in!

A birder who has progressed beyond the minimum feeder-watcher stage, seeks to know not only the name of the bird but something about its habits and habitats. Like many other recreational pursuits, birding is comprised of participants at every level of competence. They range from *early birders,* just becoming interested in birds, to *avid, keen, dedicated, indefatigable* people who have seen and identified hundreds—upwards of 800—birds in North America. They also are referred to as "those crazy bird watchers."

SEASONS IN THE SUN

Where to bird is relatively easy in North America. Many place-oriented, bird-finding guides are available. Periodicals catering to the birding public frequently feature special birding places. Some guides cover very large areas: Canada, Mexico, or east/west of the Mississippi. Some cover a province or state. Others pinpoint a very small area of special interest: Washington, DC or east-central Vermont.

When to go birding often poses sticky problems if you are look-ing for particular birds rather than just combining a speaking engage-ment in St. Louis with seeing whatever birds are around at the time. Opportunities to bird beyond home territory may open up at unex-pected times. For instance, your professional organization chooses Houston for its annual winter meeting. Should you take your binocu-lars to Texas? (Yes!) Your English birding friend must take his holiday before the end of April or lose it. Where do you suggest he begin his North American bird list?

- Where should I go for the Christmas school break to do some birding?
- Where can I see those great congregations of migrants in the fall?

- How is the birding in the Grand Canyon during the summertime? That is the only time I can really get away and, of course, the kids will be with us.

- I have a conference in Phoenix in April. What are the chances of doing some birding while I am there? I've never been to Arizona before.

- We've planned a trip across the country in May to break in the new RV and we would like to see as many birds as possible. Our time is pretty loose so we can go where the birds are.

- I am planning my first trip to the states this winter and I *really* want to bird in a *warm* climate.

British birders have seen America on the silver screen and have a historic hankering to see how the colonies are doing. Probably the most enthusiastic in the world, the British generally have a keen sense of timing; they know how important it is to be at the right place at the right time. They have been at some of the greatest migration points in the world just in time to see thousands, maybe millions of birds flying north or south: Istanbul, . . . Gibralter, . . . Falsterbo, Sweden. The Isle of Britain is a migration chute as well. Birds use the island as their pathway between Arctic breeding grounds and points south. North American bird migration "hot spots," like Point Pelee in Ontario and High Island in Texas, require careful timing too.

Many persons, particularly those who live in northern latitudes, think of vaction time as an occasion to visit warm, sunny climates. Birding vacationers know that tropical splendors may provide delightful birding experiences, but won't necessarily produce the particular birds they are looking for. If the Great Gray Owl tops your want-to-see list, try winter in Canada where the eager birder is urged to carry along "survival" gear. The Great Gray can be ticked off on your list in other northern locales at other times, as we will see. Early birders must choose the right combination of time and place to see the right bird.

Get-up-and-go time for many birders is governed by a calendar set by others: bosses, school boards, family, and so on. Recognizing that time of year is important to both the birder and the "birdee," this book is organized with calendar in hand. Don't however, believe everything the calendar says. For some birds, spring begins well before March 21. For others, the fall migration begins during the birder's summer. Roughly, the birding spring runs from March through May; summer, from June through August; fall, from September through November; and winter, from December through February. Most bird checklists recognize this designation of the seasons.

Time is the emphasis and organization of this book. Time of year is critical in seeing the Ross' Gull in northern Manitoba or the Blackburnian Warbler in Central Park. Although *some* birds can be seen in *some* places at any time of year, if you're looking for particular birds, or just wondering what birds you might see on a given trip, the time you travel can be crucial.

Time of year in many places is of most importance during the spring and fall migration when birds are on the move. In summer and winter, many species stay put for a period of time. From the birder's viewpoint, spring and fall probably are the most exciting times. Birds on the wing provide the primary opportunities for many birders to see them. We may not be able to visit breeding birds in the high Arctic or travel to Central or South America to see some woodland species that spend most of their lives there. As migratory birds pass through, often traversing the length of the continent, there may be the golden opportunity to see that Lesser Golden Plover.

Although *place* is the common emphasis in reference material, some calendar information often is provided, although it may be time-consuming to ferret out. You can start with your field guide. It provides distribution maps indicating, in a somewhat generalized way, where you will see certain birds at what time of year. Color-coded maps will indicate breeding range (summer), resting range (winter), and year-round range. Some indication of migration route (spring and fall) is often included. If breeding range and wintering range are separated by half the continent, migrants can often be found at appropriate touch-down places inbetween.

"Where-to-find" guides will give some information on best times of year. Guide books known collectively as the "Lane guides," originated by expert birder, the late James Lane, provide excellent information about birding in selected places in North America and all feature a distribution calendar. Serious birders will seek out Lane guides when traveling to a major birding "hot spot," and will give particular attention to that calendar chart in the back of each guide which indicates by month the abundance of birds.

Another major source of information about where birds are during a particular time of year is *American Birds,* the quarterly journal published by the National Audubon Society. Its reports of major bird species sightings by geographic areas of North America is invaluable. Each of the four seasonal issues reports on birds seen during the season six months earlier.

Bird lists are another quick and easy source of information. A bird list, or checklist, is a list of species, arranged in taxonomic order with seasonal information and indication on abundance generally provided (See Appendix 1). Most parks and wildlife refuges have handy bird lists available. Some are up-to-date and accurate but others may list many "theoretical" occurrences. State and local Audubon Societies or bird clubs have relevant lists covering their areas. Obtaining bird lists before you embark on a birding trip gives an idea of species likely to be seen where you're going, at the time of year you will be there. Local and regional publications are available for most major birding areas.

For advanced birders, specialized literature is available such as the *Distributional Checklist of North American Birds* (see Appendix 2 at the end of this book) that list the seasonal availability of species by state. Digging out the information in such publications can be tedious but rewarding if you want to know the likelihood of seeing a particular bird in a particular state at the time you plan to be there.

MIGRATING MIRACLES

Spring and fall are the favorite seasons for many birders, when they see more birds in the yard or in the local park. Warblers are moving through one day; the next day they're gone. Drive along the Gulf coast to catch the spring arrival of songbirds. Join a hawk watch in the fall. Warblers and other songbirds, waterfowl, and shorebirds are eager travelers. If you want to witness the migration, you must travel at the right time of year to places where you are most likely to see those avian wanderers.

Bird migration has fascinated observers since the beginning of humankind. Humans have recognized that birds do come and go in spring and fall. In Biblical times, the Prophet Jeremiah noted, "The kite in the air hath known her time: the turtle, and the swallow, and the stork have observed the time of their coming. ." Classical writers Homer, Aristotle, and others were observers of the biennial passage of birds. England's early bird observer, Gilbert White, (*The Natural History of Selborne*) wondered where swallows went in winter, speculating as did others in the late 1700s, and for centuries before, that perhaps they hibernated. Modern-day swallow watchers certainly have had a favorite place—the lovely mission near the California coast, San Juan Capistrano. Cliff swallows traditionally are expected there at the end of their journey from Argentina every March 19, plus or minus a few days. With rapid urbanization, this pattern may not be holding.

Poets have commented on the birds of passage, and scientists have studied the secrets of these small creatures. Around the world, watchers of birds mark the arrival of the first bird of spring and the departure of the last bird of fall. Often, migrating species will arrive and depart from the same place each year within a day or two of the expected date. Gilbert White kept records. My birding friends in New England have kept arrival and departure records for nearly 50 years. Some Lane guides list dates of first and last arrivals and departures.

"The Four Seasons fill the measure of the year, . . ." poet Keats wrote. So are the seasons the measure of this book. Seasonal chapters will guide the reader, in a general way, to the best times to see the best birds in the best places on this magnificent continent.

Spring is where the birding calendar begins. Julian, Chinese, Hebrew, Aztec, and Gregorian calendars aside, "nature's year begins with the spring," as Edwin Way Teale, the great American naturalist, aptly noted. Not only is it nearly everyone's favorite time, most every place in the world, it is generally the best birding time. Spring, of course, occurs at different times of year depending upon which hemisphere you're in. Spring in North America officially begins on March 21 but, for the purposes of this book, spring begins "roughly" in the month of March. As birders are aware, birds have their own ideas of precisely when spring begins. Spring defined by birders as migration and mating time, begins at different places on the continent at different times. In the British Isles, extending approximately from the 50th to the 60th latitude, it is possible to give more precise calendar guidance to the birder. Nigel Redman, in his fine book, *Birdwatching in Britain,* is able to say that in April, "The departure of winter visitors is an inconspicuous affair—they just seem to disappear."

In North America, spanning nearly half a globe of latitude, some birds are departing somewhere, others are arriving somewhere. Whenever and wherever spring occurs, birds are singing to establish territories, attract mates, and, in so doing, gain our attention. In many places, this is the best time to see migrating birds passing through. British birders call them "birds of passage," a particularly nice description of the avian rite of spring (or fall, for that matter). Some birds that reside year round in particular places are easier seen in spring when they are mating and singing, than at other times of the year.

Summer generally is the favored vacation time for humankind. Often, it is often a "hot time" for birding: birds are busy building nests and raising families. Cheeping sounds of baby birds coming from hidden nests are one of summer's listening delights. Fortunately for calendar-bound birders, "summer" for some of us is "spring" to birds in some parts of the globe. In early summer, calendar-time, birders head to where birds' reproduction time clocks begin even though we must don winter clothes as we head for northern Alaska and Canada.

Autumn attracts both "leaf peepers" and birders intent on watching the miracle of migrating birds heading back south. Hawk watching is a favored activity of many birders in the fall. Birds may not take the same route they traveled in the spring although good spring "hot spots" are often also "cool" places in the fall for a "fall-out." Birders report that "the birds were just falling out of the sky," if they are lucky enough to be where bad weather forces migrating birds down to land, sometimes in very large numbers.

Winter is when one of the most famous birding activities in the world takes place. Involving birders from all over North America, it is the annual Christmas Bird Count (CBC) organized and sponsored by the National Audubon Society. This mass event occurs in both sunny and chilly climates.

In winter, northern birders may opt for snuggling, instead of struggling through the snow, but for Sunbelt birders, winter is a favored time to be outside. Shorebirds that breed in the high Arctic, winter in Sunbelts around the world. Many wing southward in the New World and are found on beaches from the southern United States to South America. It's not as easy to sort out Sandpipers in the winter as it may be on Arctic breeding grounds. Winter, however, is when birders most often encounter the drab denizens of the shorelines. Here's where birding becomes a real challenge.

Year-round birding is a distinct and delightful prospect in a number of places in North America favored with a substantial resident bird population, location along a major migration route, and attractive to vagrants and rarities. In addition, the weather is often superb. There are birders who, if they could, would live where birding is good year-round. Many do.

Seasonal awareness is important for the birder for several reasons. Birds will be in different places, at different seasons, doing different things. Many look different at different times of the year. Nondescript, winter-plumaged songbirds don feathered finery in spring and summer. Once nesting is completed, molting commences and birds may look forlorn and bedraggled. Juvenile shorebirds require special attention. That Western Sandpiper you saw on a Gulf coast beach in August may not have been "still in breeding plumage," but perhaps a youngster in juvenile plumage. Shorebirds appear to have a plumage for every occasion. This is where studious watchers are really challenged. It is best to get a detailed book such as *Shorebirds* by Hayman, Marchant, and Prater. Many shorebirds are shown in four or five combinations of costumes: adult prebreeding, adult breeding, moulting adult, postbreeding adult, adult nonbreeding, juvenile, and moulting juvenile.

PEOPLE WHO BIRD

All kinds of people bird: early birders proud to know the difference between a Western Bluebird and a Mountain Bluebird; avid birders checking the worn plumage on molting Bobolinks; and birders at every stage in between. Early birders may eagerly join the local Audubon Society or bird club; avid birders aspire to the "700 Club" or even the "600 Club"—terms that identify birders who have attained the upper reaches of their North American "life list." They have seen more than 700 species of birds. By the time you read this, there may even be an "800 Club," so exclusive that it likely will have only one member.

You may not consider yourself "highly committed" to birding, but *Time* magazine, in its May 28, 1987, story on the growing popularity of the sport of birding, identified two million North Americans who were. The *Time* writer believes that a large percentage of highly committed birders own at least one bird guide; he reported that 600,000 birding guides are sold every year.

In the spring of 1987, some highly committed birders were visiting in south Florida.

"Message Centers" may include date and detail of location of the bird; this one is at the site of the Key West Quail Dove

"Twitched it," wrote some Cheshire, England, birders, 29th April, '87.

"You people from New York shouldn't have left."

"Rousted bird which stayed put. Rt. side of road towards water at beach. Was on rt. side 5′-10′ in—it walked across road—flew into deep mangroves on left."

These comments came from some highly committed birders who had driven long distances or flown in, rented cars, driven down a deserted road in the Florida Keys, and thrashed about in dense undergrowth near mosquito-infested mangroves to look for the Key West Quail Dove. It had, at the time, been sighted in North America only five times in this century. The birders' messages were written on paper bags near the site of their triumph.

Yes, many birders consider themselves highly committed. Wind, rain, mosquitoes, or black flies do not deter them. They believe birding is fun. Otherwise, why do it? Highly committed birders like to travel, and they have allocated time and resources to do so. Birders generally are a helpful lot, too. They leave messages telling others where to find the rare bird. When *you* get to the identified spot, look around for clues where the bird has been spotted. A couple of brown paper bags or a piece of paper weighted down with a stone may bear useful information. Near Brownsville, Texas, in February 1987, red yarn was tied to trees

Key West Quail Dove (Courtesy Jack Holmes)

where a Golden-crowned Warbler was sighted. In Toronto, a green leaf bag pointed to the site of a Boreal Owl at Humber Arboretum.

Birders are curious, studious, and patient in varying degrees. Many think little of spending an hour peering through a scope sorting out wintering sandpipers on a sandbar.

> "I'm sure that's a Western to the right, and those must be Dunlins on the left, or are they Knots?"
> "I can't see the white rump, but I *think* that's a White-rumped Sandpiper at the side of the pond."

When these birders get home, they'll look through several field guides and the *Shorebirds* identification guide before concluding that their identification was correct.

Birders really care about the objects of their endeavors. They care about protecting the rain forests, the wetlands, and the special habitats favored by special species. They take personal responsibility for their actions. Although they may scan a stinking garbage dump for a Thayer's Gull, they would never think of dropping an empty beer can out the car window.

Birders come in many costumes: old and young, tall and short, men and women (in Britain, men predominate), student and retired, white-collar and blue-collar, enthusiastic and nonchalant, beginner and expert, local expert and world traveler. They come from every walk of life. Norm Chesterfield, who placed second on the world birder list in 1988 and topped the Canada list that same year, is a mink rancher in Ontario. Benton Basham, one of the top North American birders, is an anesthesiologist. Dr. Robert Kelley, president of Tropical Audubon Society, is a mathematics professor. Professors of one thing or another are well represented in the birding fraternity. In their "other" lives, birders are realtors, actuaries, ecologists, geographers, astronomers, electricians, dental hygienists, artists, newspapermen, secretaries, businesspersons, computer specialists, and you-name-it. Many birders work at pressure-cooker jobs in big cities. They have taken the advice *Travel & Leisure* magazine gave its city subscribers a few years ago: they may find refreshment in bird watching.

One way of categorizing the birding hierarchy is to put the world birders at the top of the pile. There is only a handful, but a rapidly growing handful, in the world birding fraternity. The American Birding Association (ABA), which provides an annual listing service for those birder members who want to know where they stand in relation to others of similar persuasion, for 1988, listed 289 birders who had passed the world-listing threshold of 1200 species, over twice as many as in 1981. Such birders increasingly look upon the birding activity as a sport.

Aside from the world birders, there are many more birders who are happy searching North America for birds. In 1988, over 954 of those highly committed birders were listed as having reported seeing more than 500 species. Benton Basham saw a record 703 birds in North America in one year! More recently, in 1987, Sanford Komito of Fair Lawn, New Jersey, set a new record of 726 North American species that year. Now those guys are pretty highly committed birders. Records such as that take real dedication, hard work, and knowing when to be where.

The vast numbers of mainstream birders fall between the "hot shot" birders in the 600-plus category, and the seven million "fairly interested" birders defined by a *Time* magazine writer as those who can identify at least 40 species. How many species can you identify? When you begin to think about it, you'll probably be surprised. Quickly you've graduated yourself from the fairly interested category to the highly committed.

John Leo, in the *Time* magazine story, described highly committed birders as those who (1) watch birds regularly, (2) use a field guide, (3) keep a life list, and (4) can identify a hundred or more species. Most readers of this book will nod vigorously at the first criteria. They will also identify not one but several field guides among their bird book libraries. If you fit this profile, you may or may not keep a life list from which you can tell how many species you've seen.

Whether the estimate in *Time* of two million birders is accurate, nobody really knows. What we do know is that nearly a half million birders, environmentalists, and naturalists belong to the National Audubon Society and regularly receive *Audubon* magazine. The more "birdy" of the members may also subscribe to *American Birds,* a journal that carries in-depth birding articles and keeps track of bird sightings all over North America.

Bird Watcher's Digest has been gathering faithful readers for quite a few years and currently has over 65,000 subscribers. *Living Bird,* published by the Cornell University Laboratory of Ornithology, now reaches approximately 11,000 members. Another indication of the growth of interest in birding was the appearance in 1987 of two new periodicals aimed at this audience: *Birder's World,* "the magazine for bird enthusiasts"; and *Wild Bird,* "Your Guide to Birding at Its Best."

Really gung-ho birders, 7000 of them, have joined the American Birding Association, sort of a "professional" organization of birders. They receive the bimonthly magazine *Birding* and pore over articles designed to help them distinguish differences between an Acadian Flycatcher (*Empidonax virescens*), and a Yellow-bellied Flycatcher (*Empidonax flaviventris*). The "empies" are often difficult to tell apart.

How well do real world birders fit Leo's criteria for the highly committed epithet? A small group of North American birders with varying levels of birding interest, met each other on a birding trip to Churchill on Hudson Bay in northern Manitoba. They claimed to have been birders for between 2 and 27 years. How many birds did they think they could identify? Their answers ranged from "approximately" 200 to a more precise 475; the average among the five group members was 315. All were well over Leo's floor of 100.

Do they keep a life list? Answers ranged from "not really" to a firm "yes." Do they use a field guide? These birders not only use a field guide, but they all owned more than one. All had at least two North American field guides and two owned guides to places far afield: Panama, Mexico, West Indies, Mexico and Central America, Trinidad and Tobago, Britain and Europe, Galapagos Islands, being the favorites. Do they watch birds regularly? You bet. They watch at every opportunity—in their garden or yard, and in birding sanctuaries both near and far. Yes, this group of birders qualifies as highly committed.

As is often true with sports and hobbies, people who bird share another characteristic: They develop their own lingo. For the early birder, it helps to know what your birding companions are talking about.

"That's a lifer for me—the one that just dropped. No, it's back up —it's at two o'clock in the big tree; right in the window. See it? What a tick! I'll make the 700 club by the end of the year."

"Watch that kettle closely. Look at the one at the far right of the Broads; it looks different."

"What do you mean, the trogon is a trash bird!"

"All I can tell you at this distance is that its a Myiarchus. Could be a Dusky-capped, but more likely an Ash-throated."

"Of course the Pyrrhuloxia is related to the Northern Cardinal. In fact, it's the same genus."

Take a moment to scan Appendix 1 for a quick lesson in the committed birder's vocabulary.

TO LIST OR NOT TO LIST

If you are an early birder, you may be confused by the frequent reference to lists, life lists, checklists, and the like. Many birders pursue birding for years without knowing what a life list is. "How many birds have you seen?" The inquirer isn't referring to numbers of Mourning Doves or Blue Jays. "What are you up to now?" These common questions relate to numbers of species seen. The answers are likely to be of two sorts. "Oh, I don't know—probably around three or four hundred." This is likely to be the answer supplied by most birders. They know "about" how many bird species they've seen and don't care to spend the time on bookkeeping.

Birders who take their sport seriously generally know their "handicap." Seldom caring much about "birdies," they do know how many bird species they have ever seen, and often where they saw a species for the first time. The second answer is, "I got my 542d yesterday—a Black-chinned Sparrow." Fervent birders who hit the 700 mark may break open a bottle of champagne and be recognized in publications that are read by other birders who would be envious, or at least impressed. The *NARBA Monthly Newsletter,* published by the North American Rare Bird Alert, announced in the July 1988 edition that Bill Rydell of Las Vegas, Nevada, picked up his 700th species in Florida: a Bahama Mockingbird. For Daphne Gemmill, her 700th was a Great Gray Owl and this news was broadcast at the banquet of the 1988 American Birding Association convention. How do they know? They keep a list.

Birders are likely to keep some kind of record of species seen, perhaps in their field guide, maybe just on scraps of paper, or in some kind of formal book or booklet. Many birders routinely pick up a checklist of the national park or wildlife refuge they are visiting, but either may not check off species seen or just count up the day's total. You may be a veteran lister, but you probably remember the time when

Birders use handy checklists to keep track of species seen while visiting wildlife areas (Courtesy Jack Holmes)

you totaled the list for the first time: you were surprised at how many bird species you checked off. Making up that first list was like making up your first net worth financial statement: You're pleasantly surprised at how much money you are worth.

For serious birders, most of whom are competitive, the length of their "seen" list of species and of those yet to see, is an important enticement to travel. If you want to see—add to your list—a Purple Sandpiper, you will look along the northeast coastal areas; for a Hooded Oriole, in parts of the southwest; for a Sooty Tern, take a trip on the ocean off Florida or be where they nest. This book focuses on North America, the geography of which is covered in the next chapter. The North American checklist is generally considered to be the list of bird species found north of the Mexican boundary. Numbers of species on the list change from time to time. Currently, it has passed the 850 mark. For many birders, this is their life list.

For other birders who have strayed beyond the boundaries of the continent, their list of species seen around the world may be their life list. To put listing into a geographic framework, you could start at the top with a world list, a rather large listing of the 8600 to 9000 species of birds in the world. These and other printed checklists are listed in Appendix 2 at the end of this book. Thus, if a world birder friend says he or she has seen 2000 species of birds, you know your friend has visited places on several continents in order to amass such a total.

Next would come the North American list. Beyond that, listing categories seemingly are endless for the lusty lister: state or provincial lists, county lists, local park lists, annual lists, trip lists, and so on.

"Big Year" lists can make birding history as Stu Stuller detailed in the sports column in *The Atlantic Magazine's* May, 1989 issue. "Birding by the Numbers" tells how Benton Basham, Sandy Komito, Steve Perry, Kenn Kaufman, and others have played the game.

The listing mania sometimes goes pretty far. Bruce Neville keeps a list of "birds seen on the way to work." People keep lists of birds seen on T-shirts, on stamps, and on Christmas cards. The possibilities are endless. A favored list many birders keep is of species seen in their backyard. Backyard bird watchers living along a migratory flyway can amass impressive totals on their yard lists. Living near water areas can bring a mix of both land and water birds. Harry and Ginny Hokanson, in their first year living adjacent to Biscayne Bay south of the Miami metropolis, have a yard list including species that birders travel long distances to see. They have clocked both Mangrove and Yellow-billed Cuckoos, Bald Eagle, White-crowned Pigeon, Black-whiskered Vireo, and Northern Waterthrush, not to mention hawks circling in thermals above their roof garden.

Sometimes the listing passion just wears out. Veteran lister, author, and birding tour leader, Arnold Small, says: "Chasing around the world after 6,000 birds is not as appealing to me as it once was but I'll make it next year without really trying. The numbers game means nothing to me anymore; I just like to see new birds and new places."

If you don't keep some kind of list, try it. Especially, keep a North American list. You will probably find it's fun. To be "legal" in the listing arena, be sure the bird is alive, that it belongs in North America (or that it flew in under its own wing-power), and that it's not in a cage. That lets out zoos, unless the bird you see is a Golden Eagle flying over the San Diego Wild Animal Park or a House Sparrow helping itself to spilled feed. Those are the general "rules" birders abide by.

DO IT ALONE OR WITH A GROUP

Birding is one of the few sports, or hobbies, that can be engaged in completely solo, in an organized group, or anywhere in between. Traveling birders do it both ways. Time, circumstance, and personal preference will likely dictate which. Reputable birding tour operators have reputable leaders who have cased the reputable places. They know the best times of the year to see which species. If you wonder why birding trips are offered to Cape May, New Jersey, in the third week of September, or Canada's Point Pelee in mid-May, or the Dry Tortugas at the far end of the Florida Keys in the cusp of April and May, it's because large numbers of special birds will be found there.

On an organized birding tour, in addition to having the advantages of experienced planning of time and place, you will be able to pick the brains of some of the best birders in the country, many of whom lead birding groups. Leaders will do their best to make sure you see the bird the group is seeing, and that you note the yellowish legs of the Least Sandpiper, or the black legs of the Semipalmated Sandpiper. They will bring relative beginners up to speed. They will patiently answer endless questions. They often have superb optics such as a Questar telescope for group members to use. "I've got it in the Questar," is a joyous announcement, the signal for everyone to line up for a Hoary Redpoll. Traveling with an organized birding group pretty well assures you of getting a substantial list of birds for your efforts, the largest number in the shortest time.

In addition to the expertise of your leader, on an organized birding tour, you have the opportunity to meet other birders, often from all over the continent or even from across the ocean. There is good companionship along with the good birding and you may find a lifelong friend. Birders generally get along well with other birders, and if your group develops the best of the group qualities, the trip generates warm, good fellowship. One of the big reasons for becoming part of an organized birding group is to become part of the fun. For a Nature Travel Service group in Churchill in June 1987, it was Jim Griffith from Toronto who bonded the members of the group with his laughter-provoking antics. During those few times when birding turned dull, Jim would retrieve from a dump pile a child's bike, an automobile seat, or a construction worker's hard hat. Such treasures became props for some occasionally bawdy playacting that kept up the hilarity level throughout the trip. Think about the uses to which a hard hat could be put!

Birding on your own takes research and planning. Some birders prefer to do it this way. Check reference books for nearby state and national parks, national wildlife sanctuaries, private nature sanctuaries, and other areas noted for wildlife. Write for information and especially ask for a bird list. Traveling alone, or with a birding companion or two, is generally easy in North America unless you plan to head for the remote Arctic areas. Armed with a where-to-find guide and with some research on the time of year when the birds you want to see will be where you're going, you will do well.

Canadian birder Jim Griffith retrieves a treasure from a trash pile to amuse members of a birding group in Churchill

Trips during which birding is the primary purpose, or even one engaged in as a welcome respite from some other purpose, can be enhanced by such planning. However, if you're the kind of person who just likes to "take off," there are still loads of opportunities for birding. Be sure to have good maps; official state highway maps are excellent. Plan to picnic as often as possible, not just lunches but breakfasts and dinners, too. Early and late are the best times to see most birds. Choose your meal places by the map not the menu. If there's a state or provincial park, national park, or nature reserve nearby, it is usually worth whatever small fee is involved to surround yourself with an interesting setting.

Two

From Sea to Shining Sea

*T*his earth's five major continents are so huge that each provides habitat diversity and, as a consequence, a diversity of bird life. North American birders who keep life lists usually mean their North American list. (Peripatetic world birders think of their life list as their list of species seen around the world.) The American Birding Association in the 1986 edition of the *A.B.A. Checklist* lists 858 species. This may seem like a lot or a little, depending upon your perspective. Actually, this number of species is not large in comparison to other continents. The North American list pales besides South America, home to roughly 3000 species, about a third of the world's total. Asia and Africa also have many more species than has North America. The North American total compares roughly with the number in Europe, Australia, or bird-rich New Guinea.

To actually see North America's birds in their own habitat is to behold, along the way, some of the greatest scenery in the world. From jagged glaciers of the arctic north to arid deserts of the west; from hardwood forests of the east to vibrant flowering plants of the subtropics; there are birding experiences to tantalize and satisfy all those who would open their eyes and ears to their surroundings.

ARCTIC TO ARUBA

North America; where does it begin and end? That's an easier question to ask than to answer. Glance at the globe and it appears that continental North America extends from glacial Alaska, a long spit from Siberia, to subtropical south Florida and, perhaps, to some point south of the Mexican border. Zoologists consider the narrow neck of southern Mexico between the Gulf of Tehuantepec and the Bay of Campeche to be the southern limit of the North American continent.

Careful examination of the northern limits brings the eye near the North Pole and the Arctic Ocean. Northern Canada's Queen Elizabeth Islands lie well above the Arctic Circle. Ellesmere Island is only about 500 miles from the North Pole. These islands hug the northwest coast of Greenland. North America doesn't seem to include that icy island, or does it?

The North American continent, measured from Attu (believe it or not, a really "exotic" birding spot) on the far western end of Alaska's Aleutian Islands across to St. John's at the eastern tip of Newfoundland, would be over 5000 miles in length. Working southward below the border between Canada and the United States (the longest unarmed border in the world), the continent stretches south to . . . where?

Subtropical south Florida, is just across the Atlantic Ocean from Morocco. The Tropic of Cancer bisects Mexico from about Mazatlán to Tampico. From there south, it's tropical. Cuba and the rest of the West Indies (including the island of Aruba) hang off the tip of Florida, circling around the Caribbean Sea and then down to northern Venezuela. What about these islands?

Does North America stop at the Gulf of Mexico and the Mexican border, or farther south? What about Baja California, that tail of Mexican land hanging off the state of California? Is Hawaii, the 50th of the United States, considered part of North America? Answers to these questions depend upon who you are talking to. Even geographers don't always agree on continental boundary lines. This, however, is a book about birds, not geography. The picture of North America does need a frame, some recognized boundaries. Birds "care" about continental boundaries. Meetings of lands and seas help shape their behavior. Some birds navigate by coastal boundaries during migration. Food sources—the plants, animals, and insects on which they depend for survival—are determined in part by continental boundaries.

ZERO IN ON ZOOGEOGRAPHY

So that we're all on the same latitudinal and longitudinal length, a definition of the geography covered by this book is important. In this chapter, we try to define what we mean by North America as far as birding is concerned. There are several definitions. Turning first to the *zoogeographers,* we encounter persons who study the distribution of animals in groupings characteristic of the environment in various large regions of the world. Zoogeography divides the world into seven or eight faunal regions:

Nearctic	North America, including Greenland, down to about mid-Mexico
Neotropical	South America north to mid-Mexico including the West Indies
Palearctic	Europe, Asia north of the Himalayas, and North Africa down to the southern border of the Sahara Desert
Holarctic	Nearctic and Palearctic combined
Ethiopian	Africa south of the Sahara including Madagascar
Oriental	India and Southeast Asia, to Bali and Sulawesi
Australasian	Australia, New Guinea, and New Zealand
Oceanic	South Pacific Islands

Faunal, or zoogeographic, regions are major areas of the world where plant and animal species are more closely related to each other than they are to species in other faunal regions. Factors delineating faunal regions have to do with ocean barriers, amount of rain, air temperature, or major topographic features such as mountains. Even if you're not in the world birding class, it's not a bad idea to become familiar with this terminology. If you use the Clements world bird list to keep track of your sightings, you will have encountered these terms.

Oddly, the term *Nearctic* isn't used much among North American birders. It's not even used in introductory sections to North American field guides. The Clements world bird list doesn't use it often. That checklist is more apt to describe the range of the Yellow-rumped Warbler as "widespread North America," but that of the Meadow Pipit as "tundra and grasslands of western Palearctic region." British birders commonly refer to birds of the "western Palearctic," meaning Europe, the Middle East, and northern Africa. This term excludes the better part of Asia. Some shorebirds, waterfowl, and birds of prey familiar to North American birders do not "belong" to this continent. The Whimbrel is Holarctic, the Dunlin inhabits the Holarctic, Palearctic, Ethiopian, and Oriental faunal regions; Tundra Swans and Canada Geese are Holarctic in their breeding areas; and *our* Northern Harrier is found on the Nearctic, Palearctic, and Oriental mainland.

The better part of North America is thus the Nearctic region. Is Greenland part of North America? It is when it comes to faunal regions. No, the North American continent does not stop at the Mexican border. The Hawaiian Islands, Cuba, and other Caribbean islands are not currently considered part of the North American faunal region. For birding purposes, the zoogeographic description is helpful.

Birding organizations have different definitions. They define it for birders, not birds. The American Ornithologists' Union (AOU), that provides the authoritative definition (and taxonomic order) followed by most bird guides, has a slightly different version. Until 1983, the AOU checklist of North American species included Greenland in its definition of North America, delineated as that area north of the Mexican border. The 1983 version, honored by modern bird guides, deletes Greenland, but extends the definition of the continent to include all of Mexico and Central America, including Panama. Included are the Hawaiian Islands, Bermuda, and the West Indies except for islands adjacent to South America.

Members of the American Birding Association (ABA), who use the *A.B.A. Checklist* which covers that organization's definition of North America, count their sightings of birds occurring north of the Mexican border. The ABA excludes the Caribbean and Hawaiian Islands at this writing, but some effort is being made to include the Hawaiian Islands in the definition. This is, of course, just a definitional issue; it means not a tweet to the birds.

There you have it. For birders, the answer to the question, "What is North America?" is a bit fuzzy. For the vast majority of watchers of birds who do not keep a list of the birds they see, the distinction is not particularly important. However, most listers are scrupulous about what birds they tick off on what list. Many listers use the *A.B.A. Checklist.* For those who wish to use the broader definition of North America, and this may include birders from across the seas, the Swift *Checklist of the Birds of North America* is convenient. Both are listed in the References at the end of this book.

With this geography lesson as background, a policy of nonrigidity has been adopted for this book. Species mostly found, at some time of the year, north of the Mexican border have the right-of-way. However, as tourists and a trogon or two readily travel across the border without passport, so will this book make occasional forays into that sunny land south of the border.

Travelers find the Hawaiian Islands both reachable and rewarding, so brief mention will be made of birding there despite the vast intervening stretch of Pacific Ocean. Birders and travelers also have easy access from eastern seaboard cities to the Bahamas where birding opportunities abound when much of the continent is shivering. That's enough reason for an occasional reference to birding in these lovely vacation islands.

LIFE ZONING

Aside from a general idea of geographic boundaries, there are other geographic features of the continent that affect both the place and time birds are to be found. Life zones, or biomes, are the naturalist's terminology that describe places where we find birds we are looking for. Major vegetational associations in North America are often classified into life zones. From north to south, or from higher elevation to lower, life zones are:

Arctic
Hudsonian
Canadian
Transition
Upper Austral (east) or Sonoran (west)
Lower Austral/Sonoran
Subtropical

Such terminology may even be part of the bird's name. Consider the Arctic Tern and the Hudsonian Godwit. These zones reflect differences in temperature and humidity that affect plant life and therefore bird life. Each life zone can be thought of in terms of either latitude or elevation. On the map, each life zone represents approximately 400 miles of north-south distance in North America. Each zone supports slightly different biotic communities. As you move from one life zone to another, notice differences in the flora and fauna.

As you climb a mountain or move from low to high elevation within a short distance, you will observe the same kinds of changes as if you were traveling latitudinally. Roughly for every 2500 feet in elevation, you will move into another zone. As you climb the mountain, you will start out in the mixed forest at the bottom, move upwards to the evergreen forests, and ultimately you'll be above the tree line.

In some parts of the country, several life zones can be observed in a relatively small area. Just within the southwest New Mexico/southeast Arizona area, famous for its variety of bird life, all life zones except the Arctic can be experienced. When you start out in the desert, you are in the Lower Sonoran zone. It is hot and dry with cactus and desert wildlife. As you move higher, there is more rain, it's cooler, and pinyon and juniper trees abound. Higher and cooler yet is the Transition zone with its ponderosa pine and Gambel's oak. Above is the Canadian zone with Douglas fir and aspen. In the Mogollon Mountains of Arizona above 9000 feet, is the Hudsonian zone. It is cold, wet, and in winter, snow falls on the Englemann and blue spruces. Bird life reflects this diversity of habitat.

BASHING BIOMES

Biotic communities (or biomes) as definitions of habitat are still more explicit. They define the dominant plant life on this continent important to birds that need a special diet for survival. Some bird guides list the biomes and show them on a color-coded map. Many North American birds are found only in particular biotic communities. Birders seeking to know the birds of North America must pay close attention to the places where particular birds are to be found as well as to the time of year. The White-headed Woodpecker is easily spotted in open, coniferous forests above 4000 feet, from Washington to southern California and rarely elsewhere. Downy Woodpeckers on the other hand are rather widespread throughout the United States and Canada. A brief description of these biotic communities will help birders better understand the need to travel to many parts of the country to see a variety of birdlife.

Tundra

Arctic Tundra, the far northland of northern Alaska, northern Canada, and coastal Greenland, hosts nesting populations of shorebirds and waterfowl. These swampy lands, characterized by low growing vegetation, are home to only a few resident birds such as the much sought after Gyrfalcon and Snowy Owl. *Alpine Tundra* occurs high up on the mountains above timberline. It, too, is characterized by low growth. This is where you'll find the Rosy Finches.

Coniferous Forests

Western Montane Forests cover much of the Pacific coastal area southward into the Colorado Rockies down to an elevation of 7500 feet. *Boreal Forests,* or Taiga, reach from Alaska and the northern Rockies across Canada all the way to Canadian Atlantic coastal areas. *Eastern Montane Forests* are the southern extension of the Boreal Forest found principally in the higher elevations of the Appalachians. Each of these coniferous or evergreen communities is characterized by its own dominant species of trees and is home to such species of birds as Black-backed and Northern Three-toed Woodpeckers, Ruby-crowned Kinglet, Evening and Pine Grosbeaks, and Red Crossbills.

Deciduous Forest

Forests of the eastern United States are the typical habitat of this part of the continent. Woodpeckers, flycatchers, warblers, and other songbirds are attracted to this habitat as a comfortable and productive one in which to breed and nest.

Grassland

From mid-Texas northward to southern Canada, the midbelt of the continent today is blanketed with farmlands. The rolling grasslands, once extending for as far as the eye could see or the mind imagine, were and still are rich in birdlife. Here the birds don't need the shade and easy perches provided by the forests. Birdlife includes several grouse species, Sprague's Pipit, Lark Bunting, and McCown's and Chestnut-collared Longspurs.

Southwestern Oak Woodland

These are scattered areas of mixed woodlands and scattered ponderosa pines on hills and mountain slopes in the west. Watch for Nuttall's and

Strickland's Woodpeckers, Bridled Titmouse, and Black-throated Gray Warblers.

Chaparral

Although found in scattered inland areas from Oregon south to Baja California, these areas are primarily located in southern coastal California. Characterized by dense, dry scrub or brushlands along low mountain areas, this is the place to look for Wrentits, a delightful "lbj" (little brown job) known only to North America. Park near a scrubby hillside and listen for its call any time of the year. Other ground-dwelling species like California Thrashers, Green-tailed and Brown Towhees, and Black-chinned and Fox Sparrows are probable.

Pinyon-Juniper Woodland

These scattered areas in the southwestern United States are dominated by pinyon pines and junipers on hills and mountain slopes. This is where the Pinyon Jay resides along with such other species as Common Bushtits and Plain Titmice.

Sagebrush

Here's what many of us think of as the "true West": hot and dry in the summer, hilly areas largely covered with scrubby bushes and dominated by sagebrush. Much of Nevada, southern Idaho, and western Wyoming is sagebrush country. You will find the "three sages" there: Sage Grouse, Thrasher, and Sparrow.

Scrub Desert

This is as low as you can go. The lowland deserts, hot and dry, are characterized by scattered mesquite, agaves, and cactuses. Roadrunners will be running and Le Conte's and Crissal Thrashers will be thrashing about in the low vegetation. Elf Owls, Gila Woodpeckers, and Cactus Wrens nest in the tall saguaro cactus.

These brief descriptions are meant only to supply a handle on the different zones of North America that support different plant life and consequently different bird life. If you would find particular species, pay attention to what your field guide says about where they can be found. *Meadow*larks are not likely to be found in the coniferous forests, but you need to scour the scrub to find the Scrub Jay. Most characteristic species mentioned are residents of areas particularly adapted to

Cactus Wrens are named for characteristic plants of the Scrub Desert

Scrub Desert

their needs. Although it is generally easier to see most species in the spring when nesting activity is more evident, residents will be found in appropriate life zones year-round.

FLYING THE FLYWAYS

In contrast to resident species, other species migrate. They winter in one place, nest in another, sort of like having a second vacation home. It is the migrating birds that stir the blood of many birders. Traveling around North America, birders want to be in the right place, at the right time, to see these restless birds of passage.

Ornithologists prepare maps based on detailed observations that show the general time and place a species will be as it moves toward nesting grounds. As a result, we know much about distances flown and destinations of many migrants and less about how they do it. For bird watchers at all levels of expertise, from the feeder-watchers to the "700 clubers," we watch and marvel. We learn something about where different species are in the winter and in the summer. We learn the pathways birds take from winter grounds to summer breeding grounds. We watch for the particular species that are known to fly along certain pathways or flyways.

Some birds, like the Arctic Tern, travel mind-boggling distances, breeding every year in the Arctic, wintering in the sub-Antarctic. Some species take their time; others race. The *New York Times* reported on September 1, 1987, that a Semipalmated Sandpiper had flown from Massachusetts to Guyana in South America in four days, a distance of 2800 miles. Now that's fast!

Most birds fly south to north to breed. The South Polar Skua is one of the exceptions. It breeds in Antarctica, spending the rest of its year over far northern waters. A few birds fly southeasterly or southwesterly from Arctic areas to the far reaches of the western hemisphere. Some, like the Bristle-thighed Curlew are round trippers across the Pacific from Alaska to Hawaii and other South Pacific islands. The Arctic Warbler, a *Phylloscopus* Old World warbler, breeds in Alaska and makes the long journey to the Philippines for a winter vacation. Some birds migrate from high to low, an altitudinal migration occurring particularly in the boreal forest. The Clark's Nutcracker, generally found just at timberline in the Rocky Mountains, often comes down in winter to lower elevations where food is more plentiful.

North America has four major flyways, each up to several hundred miles wide in places. The Atlantic and Pacific flyways hug the continent's coastal areas. Some waterfowl and shorebirds use both coastal flyways to travel from high polar breeding areas on Ellesmere and Baffin Islands and along coastal Greenland. On their southerly journey, many species take the Atlantic route which generally follows the St. Lawrence River southward through New England, then down the coast to Florida and points south. Some birds head for Mexico, Central America, and as far south as Argentina.

Pacific coastal flyway species often have been nesting in Alaska or northern Canada and are following the coast down to Baja California, some flying on down the western coast of South America. Brants, a sea-going goose, primarily choose this route, wintering along the California coast on down to Baja although some head for the Great Lakes or down the east coast. Wandering Tattlers, Surfbirds, and Black Turnstones are among those species that wander down the west coast, the Surfbird going all the way to the southern tip of South America.

The Mississippi flyway roughly follows the river route branching off in several directions at the confluence of the Mississippi and Missouri Rivers. Flying northward, birds heading up the Mississippi River valley swing northwest across the Great Plains. St. Louis is a reference point for this midpoint of spring. Spring migration is in full swing in this midwestern United States city. Famous Forest Park with its great deciduous trees is the place to meet your birding friends. Many birds generally means many birders and the St. Louis area has both.

To the west, the Central flyway, a very broad migration route from Brownsville, Texas, up through the Panhandle, covers the better part of the midsection of Texas. Running close to the 100th meridian, it is one of the reasons Texas is such a great state for birding. This is where eastern and western birds meet, many of them migrants flying up from Latin America. The Central flyway extends northward across the grass plains to reach nesting grounds in the northern states, Canada, and Alaska. A Miamian heading westward from Austin on a summer family camping trip years ago, saw so many new birds that picnic breakfast preparations at a roadside stop turned into a gastronomic disaster. Remember seeing your first brilliantly red Hepatic Tanager? It is better than coffee to wake up the early birder.

Peter Alden, well-known world birder, points out that birds such as songbirds, migrate in broad fronts. Some larger birds migrate along relatively narrow flyways in large concentrations. Just don't expect to see the skies blackened with birds as in days of yore. It is too late for most of us to experience such spectacles.

Those aerial highways, aren't made of concrete; different species fly different routes. Some use the same route both northward and southward, while others, like many of the Lesser Golden-Plovers, fly quite different routes. Its northbound route from southern Brazil goes up through Central America to Arctic breeding grounds. It then changes its route in the fall, flying southward over Newfoundland and the Atlantic Ocean to reach its winter range.

Canada and the United States have established wildlife refuges all across North America, particularly along these aerial courses. Northerly refuges provide protected breeding sites for many species, particularly waterfowl. Southerly refuges are winter resting places. Refuges in the middle of the continent may be used either for summer nesting, for wintering areas, or may be stopping-off points in which migrating species may briefly rest and feed.

Such refuges have provided succor to millions of birds but even such massive efforts as have been made are not enough to stem the tide of declining habitat. The *New York Times* reported on February 9, 1988, a 35 percent decline in the number of Mallards in the 30 years between 1955 and 1985 and a 66 percent decline in Northern Pintails during the same period. The *Times* also reported on a massive effort by United States and Canadian governments to restore those flocks of wildfowl. Not only will this effort require cooperation of farmers along migration routes, it will require more sensitivity on the part of the weatherman. The killing heat wave and drought of the summer of 1988 seriously affected waterfowl dependent on watering holes and food sources.

In the United States, acquisition of lands comprising the National Wildlife Refuges has been funded by proceeds from *duck stamps,* the purchase of which is required of all duck hunters. Many birders also buy duck stamps which provide free entrance to national wildlife refuges.

WEATHER WISE

For birders eager to see a particular species, it is nice to know precisely when to be where to catch that ephemeral passage. Alas, such is not possible. To come close, watch the weather. In the spring, passage of a warm front seems to encourage a particularly heavy migration. But along the Gulf Coast, a cold front in early May may trigger a fallout of northbound migrants that drop exhausted from battling the winds during their flight across the Gulf of Mexico. In the fall, the migrants seem to follow behind a cold front.

American Birds, the National Audubon Society journal, emphasizes the importance of watching the weather. Writing about the spring 1986 migration in the northern Great Plains, Gordon B. Berkey notes "The early warm weather pushed migration far ahead of schedule. In all, there were 67 earliest-ever (including ties) species records for the Dakotas." In the southern part of that range, Frances Williams says,

"The wind blew steadily, mostly from the south, so one assumes the migrants went over without stopping." Of the fall 1986 migration, Ron D. Weir wrote of the Ontario Region that "Songbirds took advantage of ideal night weather during August and September to migrate southwards and most flycatchers, vireos, and warblers moved out early."

In addition to weather signals, there are other, more fixed and sure guides such as lakes and mountain ranges. Lakes are important resting and feeding areas for waterfowl moving north in the spring. Mountain ranges, particularly in eastern United States have generated well-known hawk watches in the fall.

THE GOOD NEIGHBOR POLICY

Virtually all wood warblers that cause the North American woods to ring in the spring with their cheerful songs spend their winters in Latin America. Only the Pine Warbler of eastern North America ventures no farther afield than Mexico and the West Indies. Blackpoll and Connecticut warblers winter as far south as Brazil. Wood warblers are New World warblers. Many other birds common in North America winter in Central and South America: vireos, tanagers, orioles, finches, and buntings. What North American birder on a winter visit to Mexico has not quickly recognized a White-eyed Vireo? A birder of close acquaintance spotted her first Tennessee Warbler in Guatemala, waiting six years before she could add this very common but often hard-to-see (easier to hear) warbler to her North American list. Red-eyed Vireos show up in odd places all along their migration route down as far as Argentina. Although some migrating raptors winter in parts of the continental United States, many winter in Latin America.

Strong bird ties bind the two Americas. Many species that through the ages have wintered in the south, come north to breed, some using the Central American land bridge. Southbound hawks often crowd the air over the Panamanian neck in stunning numbers. Because of these persistent geographic and biologic ties, we recognize that the very future of birding in North America is gravely dependent on what happens economically and politically with our southern neighbors. Logging of forests for commerce and agriculture can destroy the habitat necessary to support many species. Our seemingly insatiable demand for cheap hamburgers has caused irreparable damage through massive clearcutting of thousands of square miles of tropical rain forest to provide cattle grazing lands. Experts believe that between 25 and 100 acres of tropical forest are lost around the world every minute. They reiterate that when the forest disappears, so does the other flora and fauna in addition to the birds. If forest destruction in places like Central America and Brazil is not curbed, North American springs could become more and more silent.

Noted author, biologist, and writer Paul R. Ehrlich, pointed out in the *Sierra* magazine's September/October 1988 issue that the forests of Central and South America are among the "most endangered habitats on Earth." The decline in that habitat for wintering birds, according to Ehrlich, is reflected in the decline of species such as Red-eyed Vireo, Acadian Flycatcher, and Hooded Warbler. The Canadian Wildlife Service has estimated that among the species expected to be most affected by the breakneck loss of forest habitat are the Philadelphia Vireo, Orange-crowned and Palm Warblers, Yellow-bellied, Western and Great-crested Flycathers, Ruby-throated Hummingbird, Vaux's Swift, Rough-winged Swallow, Turkey Vulture, and Northern Oriole. Perhaps we need to re-member what happens in the coal mine when the canary dies!

WORLD TRAVELERS

In addition to the ties that bind North and South America, the winds that blow sometimes blow in birds from Europe, Asia, or Africa. New species reach our shores with regularity. Many birds we see in North America are common in other parts of the world. At the right times of year, they will be familiar sights to birders from other countries. Shorebirds, waders, waterfowl, and seabirds are examples. Some species are identi-cal wherever they are found. The Red-necked Phalarope, Pomarine Jaeger, and Glaucous Gull are "Arctic Circumpolar"; they breed in arctic areas and spill down the globe all around the world like paint poured on a ball.

Purple Sandpipers breed in northern climes and winter on both sides of the North Atlantic Ocean, as do Northern Pintail ducks. The Green-winged Teal will inspire a flash of recognition on both sides of the Atlantic; the American version is a subspecies of the Common or Green-winged Teal found in Europe, Africa, and Asia. The Tundra Swan which clusters in enormous numbers in favored winter watering holes after breeding high in arctic North America, is the same species as the Bewick's Swan British birders eagerly await in the fall; they are just dif-ferent races.

Pelagic birds, those birds that live their lives at sea coming to land only to nest and breed, often are common to both North America and Europe. Species such as Leach's Storm-Petrel, seen off both coasts of North America, are also seen in British and western European coastal waters. Northern Fulmars, several shearwaters, Northern Gannets, and other seabirds are familiar to both North American and British seabird watchers.

A few other birds are "at home" any place in the world, Antarctica excepted. The Osprey, common over fish-filled waterways in many parts of North America, is found all over the world except in Antarctica. It has no close relatives and is not called by different names in different parts of the world, although the species division into several races indicates special characteristics relating to geography.

THE WORLD OF BIRDERS

The world begins where you are. For North Americans, it is their own vast, beautiful continent that holds exciting birding treasures. For the British, and increasing numbers of Swedish and Finnish birders, and other "first timers," this continent offers the excitement of seeing Florida and frigatebirds, Texas and turkeys, California and Calliope Hummingbirds, Oregon and juncos, Arizona and Acorn Woodpeckers, and Saskatchewan and swans. British birders whose world is fueled by feathers, find the Atlantic Ocean not a barrier but a lure.

Arizona and Acorn Woodpeckers go together

PLACES TO BIRD

"Why don't you come along?" Your friend has just announced a birding expedition to the local park. This question has started many an incipient birder on the path to greater enjoyment of the outside world. As you learn the birds at home, you begin to thumb through your field guide and notice that a bird you would particularly like to see is "common in low mountains, lowlands, and along coasts." If you live on the prairies, you must then travel elsewhere to find this species.

In the course of traveling more to see more species, birders learn the generic places useful to explore. Special favorites include garbage dumps and sewage settling ponds, known by distinctive aromas; airports, known for their noise; and cemeteries, known for their quiet occupants. All have special attractions for certain birds. Within walking distance of the Winnipeg airport is the Brookside Cemetery just alive with arboreal birds. Airports are good places to watch, so don't pack your binoculars. While waiting at that Winnipeg airport in June, our Churchill-bound group spotted a Chestnut-collared Longspur, a life bird for some. Burrowing owls are partial to airports and golf courses.

While on the road, don't forget highway rest stops. They generally are large, clean areas well back from the highway, often completely hidden from passing traffic. Occasionally, barbeque grills and electricity encourage culinary pursuits. Picnic tables are often located within

Burrowing Owls favor airport areas

covered areas to protect their occupants from sun, wind, and rain. A grove of trees or a bit of water signal an oases for bird life. If the western landscape is devoid of trees, careful landscaping makes the rest stop even more attractive to both birds and humans. Just don't ever leave the car with your binoculars inside. Who knows, you could spot a life bird on your way to the restroom (also generally clean, well lighted, and ventilated.) State welcome stations, conveniently located near the state line on most major highways, are particularly nice places. There may be gardens, bird feeders, and nesting boxes outside, as well as maps and other information inside an attractive, occasionally historic, building.

Listers may even want to open up a "rest stop list." I must admit a considerable predilection for highway rest stops. When driving long distances, it is hard to pass up one, whether I need a "rest" or not. That blue sign with the white lettering posted along major highways throughout the Unites States is an irresistible lure. I pull in, grab my binoculars, lock the car, and walk through the often rather extensive areas to scan sky, trees, and forest or mesquite thicket beyond rest stop boundaries.

It seems that some rest stops specialize in House Sparrows and transcontinental trailer trucks. If this be the case, do your "resting" and leave. Others will have nesting birds in season, Cactus Wrens cleaning out garbage cans, Scaled Quail scuttling under some bushes, White Pelicans on the lake in the distance, a Golden Eagle over the mountain, a Black-necked Stilt screaming at some unseen intruder, Tufted Titmice flitting about the trees, Ladder-backed Woodpeckers doing their thing, and rabbits popping in and out of their burrows.

In Canada, highway rest stops are less common. Resting along the highways, however, is often the only alternative. Vast distances between cities in much of Canada leaves little choice but to take picnic fixings and pull alongside a little lake or down a side road in the woods. It will be clean and pretty. Driving through the many provincial and national parks and forest areas provides plenty of opportunity to stop at established picnic places. In fact, outside of the major metropolitan areas, they are your basic choice for lunch.

Birders on the road seek places where birds, rather than Buicks, abound. Getting off the interstate here and there provides opportunities to pass through once busy small towns. Often there is a town square worth a look; or oak trees shading the county courthouse encourage a look through the leaves.

Some birding drivers proceed like the plover: a few rapid steps and a stop, a few more steps and a stop. These birders drive and stop frequently—off the main expressway. Even though they have seen Pileated Woodpeckers before, the temptation to stop and take a look is irresistible. Bill Bouton, a Michigan birder who has seen more birds than most of us (over 700!) disagrees with this approach. "It may be fun but is not the best way to add to your list. Several carefully selected "hot spots" will add many more species than will weeks of poking around rest areas, cemeteries, and courthouses." For some, birding is a serious pursuit. Serious or casual, birders are birding because they have fun and enjoy watching birds.

Birds show up in some odd places. Traveling birders keep in mind not only common places like highway rest stops and public parks, but private wildlife sanctuaries, arboretums, and public gardens. In February 1988, some rare birds showed up in some rare places according to the North American Rare Bird Alert (NARBA) *Newsletter.* They showed up on beaches at high tide and low tide, along a jetty, at a sardine factory, dumps, picnic areas, a dam, a power plant, a sewage pond, a fire station, and a backyard feeder. Keep your eyes and ears open (if not your nose), wherever you are.

In addition to such places found worldwide, there are other places birders think of when they begin to categorize birding opportunities. Major ones include:

- *National Parks.* Areas preserved as national parks in the United States and Canada will encompass areas that attract birds. Not all national parks, however, are prime birding locations. The National Parks and Conservation Association in its publication, *National Parks,* listed six national parks in the United States it thought provided the best birding: Big Bend, Texas; Everglades, Florida; Haleakala, Maui, Hawaii; Olympic, Washington; Acadia, Maine; and Fort Jefferson, Dry Tortugas, Florida.

- *National Wildlife Refuges.* These refuges have been established in the United States and Canada to preserve specialized habitat needed by wildlife for its continued existence. Bird refuges have often been acquired and preserved to protect migrating species and thus are most interesting during spring and fall. Most refuges have bird checklists that contain some seasonal information.

- *Other Public Preserves.* National forests, seashores, grasslands, monuments, scenic riverways, and so on all offer the opportunity for birding.

- *State, Provincial, and Local Parks.* Many state, provincial, and local parks preserve areas of scenic beauty to complement the beautiful birds. Some of these parks, however, offer water skiing, archery, horseback riding, and other recreational pursuits, and may not be prime birding locations, particularly on holiday weekends and during peak vacation times.

• *Private Preserves.* Many private groups at the local, state, and national level, have preserved valuable lands that may offer fine birding opportunities. Principal among such organizations is the Nature Conservancy. Some have limited access, some have limited bird life, and some have organized tours. If you are interested in visiting any of them, it would be wise to make inquiries in advance. Directions and the source of further information are included in the Conservancy's list of 39 of their "most visitable" natural areas (*Guide to Preserves,* The Nature Conservancy, 1800 N. Kent St., Arlington, VA 22209).

The National Audubon Society also operates a number of nature preserves, a list of which may be obtained from the Society.

When traveling around the country, you will always be near some birds. In a campground, birds probably will be nearby. Even if you are staying at a motel or hotel, survey the grounds. If there are trees or a bit of water nearby, investigate. At a motel near a busy expressway in Brownsville, Texas, a pond in back was checked out before I checked out. Approaching quietly, I saw a small flock of Black-bellied Whistling Ducks. Lovely birds with their red-orange bills and gray faces marked with white eye rings. They allowed a short viewing, then flew off.

RUN FOR THE RARITIES

A few zealous birders get their jollies by stretching the limit of this continent's species. They always have hopes of spotting new ones. That desire to continually add new species to the list is what recharges birders' batteries. How else can you explain the craziness of birders who visited Everglades National Park in June 1987? Swatting mosquitoes and wiping perspiring faces, they searched sere grass and low trees for a rather nondescript black bird, the Shiny Cowbird, on one of its first visits to this southeastern tip of North America. Birders have mixed feelings about this interloper that now seems to be a regular summer visitor. They hope that it will be added to the "official" list of species in North America, another "tick" on their list. On the other hand, cowbirds parasitize other species nests and often drive off the native birds. This tick may not be a good one.

Big birder Benton Basham advises birders who would pursue birding with some seriousness to look for the "hard" birds first, suggesting that the "easy" ones will come along the way. He's developed a system that ranks birds from 1 to 5 in degree of difficulty of finding them.

Implicit in this system is that, extremely unlikely though it is, it would be far more exciting to see an Ivory-billed Woodpecker, a 5, than a Pileated Woodpecker (exciting though it looks) that ranks as a 1.

British birders are among the most eager of the rarity seekers. North American birds blown on ocean winds to the British Isles are revered as rarities. Laughing Gulls and Song Sparrows rarities? You bet, when they land on foreign shores.

So it is in this country. One April in Newburyport north of Boston, a line of cars parked along the road proclaimed a rare sighting. Scopes were set up and a couple of dozen birders were peering intently at an array of gray shorebirds in a shallow pond. Someone had spotted a Reeve, the nondescript female of the Ruff, an occasional visitor from the other side of the Atlantic Ocean. We, too, set up our scopes and tried to figure out just which of those small birds had caused all the excitement. "You see the four Willets on the left in the second bit of water from this side," offered a helpful scoper. "Well, it's just to the right and in front of them, moving left towards the dowitchers. It's fatter with a shorter bill and just a little more brown than the others. Got it?" Male Ruffs in regal breeding plumage in England do cause a bit of excitement, but in England, the Reeve gets notices rather than raves.

A rare bird, or rarity, is one that is infrequently or uncommonly seen. The Tropical Parula is a rare resident of the Rio Grande Valley in Texas. Great Gray Owls are uncommon in their range. Birders hunt hard to see such species. A rarity may also be a common bird in one part of the continent but rarely seen or accidental in another part.

Vagrants, when we're talking about birds, are those that have been blown off course or wander into a different territory. The Muscovy Duck, that has become domesticated north of the Mexican border, is found wild from Mexico to South America. When one of those wild ones shows up in Texas, it's worth a listing by the North American Rare Bird Alert.

For purposes of the phrase, *Run for the Rarities,* we're talking about those species that are uncommonly seen or have accidentally arrived on the continent or in its waters. Some such species have been added to the official North American list. Others, such as the Collared Dove, are standing by, in the wings perhaps.

Use of the term *rarities* generally does not apply to the *exotics,* those birds that belong somewhere else but were brought to this continent by human hands. Many of the latter have escaped from cages or from importers. The Red-whiskered Bulbul, native to Oriental Asia but found in south Florida, is one such escapee that has adapted to the environment and is breeding successfully. The Bulbul and the Canary-winged Parakeet are now established in the wilds of the Miami area

Sulphur-crested Cockatoo is one of many south Florida exotics

and are "legally" countable by those birders who count the number of species they have seen.

The Mexican-United States border is a bird-rich area. Some species nesting primarily south of the border care not a quill whether they invade United States territory. Stunning Elegant Trogons have found some choice nesting trees in the Chiricahua Mountains and a few other places in southeast Arizona. North American rarity seekers rushed to that area in the spring of 1985 to see the first Flame-colored Tanager recorded north of the border. A "first North American record" or a "first state record" are generally rarities worth running after. Ringed Kingfishers of Mexico are residents in the southern tip of Texas, one of several species that prods many North American birders to "run" to the lower Rio Grande Valley. The Brownsville city dump is famous among birders for the Mexican Crows that are regular winter visitors.

Coastal areas offer other attractions as birders hope and pray for a good storm to blow in some "new" birds. Birding excitement was intense in April 1983 when a Western Reef Heron, closely related to Snowy Egrets, visited Nantucket Island. It is a wading bird that belongs in West Africa and points east. West coasters also keep an eager eye out for off-course birds. Streaked Shearwaters off the Monterey

Peninsula in California are sure to elevate the blood pressure for pelagic birders as will the Wedge-rumped Storm-Petrel and the splendid Red-tailed Tropicbird.

For the North American birder bored with Western Flycatchers and Black-capped Chickadees, there is romance in spotting Siberian species, many of which touch down in far western Alaska. The moniker "Siberian" is attached to Flycatcher, Rubythroat, Blue Robin, Accentor, and Tit. These species have all found their way to North America and appear on the *A.B.A. Checklist.* Try also British Storm-Petrel or Chinese Egret. "Oriental" delineates Pratincole, Cuckoo, Scops-Owl, and Greenfinch. "Eurasian" is now used to distinguish Eurasian from related species in North America. Eurasian Kestrel and Curlew are such examples.

These species aren't common. You need a rugged determination, single-minded dedication, and a bit of luck to find them. The reason "hot shot" birders streak for Attu in the spring is not because of the salubrious climate of this island at the end of the Aleutian chain, but because Attu is just a few ocean miles from that curious land appendage of the Soviet Union which separates the Bering Sea from the Sea of Okhotsk. Siberian species are found with some regularity here and in other western Alaska spots. A regular, almost common, Siberian visitor to Alaska is the Rufous-necked Stint. That species even made it as far as the Jamaica Bay Wildlife Refuge outside of New York City in August 1985.

If you have done your homework, have an idea of birds you are likely to see, and want the excitement of the chase, you need to know about the up-to-the-minute assistance that is available. First, there may be a local rare bird alert or hotline in your community. Call an Audubon group or local bird club and ask a local expert how you can contact it. The ultimate of the rare bird alerts is the North American Rare Bird Alert (NARBA). Subscribers receive a code name and number and are able to access the national hotline to find out what rarities are currently being seen in Alaska, south Texas, or anywhere else in the country. This game can become so absorbing that you may begin dashing around the country to see the latest rarity.

Once a rarity has been sighted, various hotlines get the word and, depending on how rare the sighting is, birders may wing and walk in from all over the country, even the world, to catch a glimpse. The British call this "twitching." If the bird appears to be a "new" species deserving of serious attention, experts will take notes on plumage, behavior, and other circumstances. Photographs will be taken and videotapes made if possible. Field guides will be studied and identification agreed upon. A relatively rare species that might be expected in a particular habitat, likely would be listed in a local or national newsletter. If it's a record of some sort, all the documentation will be assembled by appropriate experts and, a year or so later, it may be written up in a journal and added to the official list.

Although you are unlikely to see rare species on your visit to Yellowstone National Park or the Blue Ridge Mountains, each place and each time of year has special treats for those who take binoculars and bird guide whenever straying from home territory.

Three

Spring:
A Time to Move North

At the top of the list stood the blooming of violets and dogwood, the return of the bluebirds, the songs of redwings and robins and the calling of the spring peepers.

Edwin Way Teale
North with the Spring

*C*ardinals had been singing for six weeks and the Screech Owl had become restless, trying out roosts in both the Sago palm and the mango tree. Drab Ruby-throated Hummingbirds had not been "clicking" around the orchid tree for several weeks. Mourning Doves had returned to the birdfeeders, safe in the knowledge that the Sharp-shinned Hawk had departed the garden. Overhead, Purple Martins had been swirling and burbling for weeks, and underfoot on coastal beaches, knots of still-gray birds gathered, exhausted from an early leg of their northern journey. Suddenly in south Florida, it was really spring, March 21, and the Spot-breasted Oriole began to sing. Soon, in the Colorado Rockies, the sun would be boring into snowy meadows, and mountain streams would break through their winter cover. In Vermont, "mud season" was not far behind. All across the southern border of North America, migrating songbirds would be arriving.

Spring, that wonderful time of rebirth. Spring may be your favorite time. Birds move north, acquire their nuptial plumage, and sing their own hallelujah chorus. From the backyard feeder-watcher to the world birder seeking new families of birds, this is the season to polish binoculars and scopes, to watch the old familiar birds, and to pursue new species.

For birders who leave their home territory to see what spring has to offer elsewhere, spring presents more choices than a field guide has plates. Will it be southeast Arizona this spring, High Island, Texas, or Point Pelee, Ontario; Big Bend or the Dry Tortugas; the Louisiana coast or Colorado; the Platte River in Nebraska or Jamaica? Birders need a calendar that will allow filling in springs for the rest of their lives.

The best place to begin your spring birding is where you are. Do you know the birds that move through your yard, your neighborhood, your nearby park? Do you know which species are year-round residents? Do you recognize the ones that you may only see one day as they stop briefly during their northward journey? Do you know the ones that arrive and stay to build their nests and raise their families? One of the most frequently asked questions from a visiting birder is, "Does Scott's Oriole breed here?" If you've done your homework, push off to another part of the continent.

WHEN IS SPRING?

Funny question? No, not really. What we think of as "spring" occurs in different parts of the globe at different times of the calendar year. Spring on the northern hemisphere birding calendar runs from about March to about May. If you would be where the birds are at any given time, developing a "bird brain" is mandatory—think like a bird. Spring for birds, regardless of calendar dates, is the time when juices begin to flow, when mates are chosen or merely rediscovered with billing and cooing, and when the urge surges and young birds soon fill the nest. Spring's the time when songbirds sing, an enormous help in finding the birds you are looking for.

Birds have their own instincts about just where to be when. Some migrants actually begin moving northward in *FEBRUARY*, others may not attend the rites of spring in the far north until *JUNE*. Although the swallows come back to Capistrano more or less on schedule and some birders claim to be able to set their watches for the time birds return to their feeders, many factors affect the precise moments when migratory birds pop up in different places.

Although naturalist Edwin Way Teale moved "north with the spring," it may not be the best strategy for the birder with a limited amount of time. The idea, however, has some merit as you think about being along the southern boundary of North America and sort of pushing the snow northward in front of you. If you can't start when they do, your migratory imagination will suggest some midsection of the continent for a visit later in the spring.

Visualize that southern part of the continent running from the southern Atlantic coastal area, along the Gulf Coast, and pushing westward along the Mexican border to the Pacific Ocean. Most anywhere along that imaginary line will put you in a position to catch the migrants coming up from Central and South America. Some will have flown long distances before they reach their nesting continent. Many species, such as Eastern Kingbird, Bobolink, Upland Sandpiper, and Hudsonian Godwit, winter entirely south of the equator. Other species that winter in the southern United States will already have moved north and are selecting nesting sites by the time you put away your winter coat.

One technique for figuring out the best time and place for finding special birds is to follow the lead of the knowledgeable birders. Write to one or more of the birding tour operators listed in Appendix 2 at the end of this book. Their catalogs will give you a sense of when they will be taking groups to special "hot spots." You may succumb to their enticing descriptions of the courtship dancing of male Greater

Prairie-Chickens in *April,* or you may have different priorities and want to go it alone. You will get a pocketful of ideas of when to be where. Check into information about national parks and wildlife refuges for general information about the best time to see the wildlife of that area. This chapter discusses some of the tips from such sources.

EARLY BIRDING

FEBRUARY is spring for shorebirds starting their restless flights north along both coasts. If you're beachcombing along south Florida beaches or patrolling southern California coastal areas, you won't notice a whole lot going on. Shorebirds that wintered in South America will be moving northward but it will not be until *April* and *May* that their migration movements will become obvious. Coastal beaches then will almost be heaving with sandpipers, turnstones, plovers, and their friends. The real shorebird extravaganza, however, will not occur until summer and you'll have to be up in Arctic reaches if you want to see it.

MARCH is the first time during the *birding* year you may see the rare and endangered Whooping Cranes if you're down along the Texas coast. The highlight of a Texas coastal trip in winter, cranes can still be sighted in very early spring at Aransas National Wildlife Refuge. They are best seen by boat so inquire at your motel in Rockport for directions. Don't wait much after mid-*MARCH*.

MARCH through early *APRIL* is what you're really waiting for: the time to attend to one of the earliest of spring spectacles luring the binocular buffs from all over the world. It is the magnificent pageant of 500,000 Sandhill Cranes, an estimated 70 to 80 percent of the world's total population, gathering along an 80-mile stretch of the Platte River, that courses through the breadth of Nebraska. (Grand Island is the reference city). This eons-old ritual observed by these stately birds offers an unparalleled experience to birders willing to brave the late winter chill. Watch the immense prancing and preening ballet performance soon after dawn and again, for the matinee crowd, in the evening. If you are in command of your calendar, try the third week in *MARCH*. With patience, you may behold some early arriving Whooping Cranes. A few will be stopping briefly here on their way from Aransas to their breeding grounds in the Wood Buffalo National Park, Alberta.

Although the cranes put on a stellar performance, theirs is not the only spectacle. A quarter million Greater White-fronted Geese and tens of thousands of ducks and Canada Geese will be in the neighborhood. If you tire of these, look for Smith's and Lapland Longspurs which may be lingering on these wintering grounds. Permanent residents such as Sharp-tailed Grouse are often found at numerous nearby wildlife refuges. The Nature Conservancy, Whooping Crane Maintenance Trust, National Audubon Society, and Rainwater Basin Wetland Management District maintain viewing areas for cranes and huge flocks of waterfowl.

Sheyenne National Grasslands in North Dakota has placed special emphasis on protecting Greater Prairie Chickens. These large game-birds put on their finest performance in the spring when they are most easily seen. Just get up before dawn and be at their courtship display grounds, or leks, while it's still dark. Many leks are well known, for the birds strut and display on the same grounds year after year. If you have not seen them at the well-known leks in Colorado, try these.

Traveling birders seeking the best spring birding should have a working knowledge of major flyways (see Chapter 2). Follow the route in spring from south to north up any of the flyways. Move up the Atlantic flyway from south Florida to the Maritime Provinces and beyond. Or try the midsection of the Mississippi flyway traveling north along that famous river. The huge Central flyway is one of the reasons Texas is such a stunning place for spring birding. It is there that the flyway is at its broadest. On the west coast, it will be the Pacific flyway that provides birding excitement as the annual ritual is played out. If decisions are your weak point, you can avoid the flyway priority and throw darts at the map. Or sally forth along a southern coastal route, sampling the early fruit of the flyways across the continent.

GULF COASTAL RIDE

APRIL is the best time of year to travel through or visit the Gulf coastal areas. It's a good time for an automobile trip to catch some wintering species still hanging around. Central and South America species heading for summer breeding grounds will delight the traveler. Follow along from Miami to High Island and other Texas "hot spots" in mid-*APRIL* 1987, and decide whether this route is for you.

Crossing Route 41, the Tamiami Trail, to Florida's west coast, I pause at the Miccosukee restaurant, as famous a stop as is the Patagonia rest stop in Arizona. Here on the Indian reservation, birders watch endangered Snail Kites flapping low over the Everglades. I see two that day. Farther west, a Swallow-tailed Kite flies over the road above the car. Long-legged waders—ibis, egrets, herons—linger in the many watery areas although most have packed and left for northern nests. Reluctantly, I pass up Ft. DeSoto Park just south of St. Petersburg, a super spot for a landfall from mid-*APRIL* to mid-*MAY* most years.

Driving northward along coastal Route 41 instead of taking the turnpike, I spot three stately Sandhill Cranes strolling along the road shoulder north of Tampa. I wonder if they are part of south Florida's year-round population or the remnants of the migrating group. Later, pulling down a side road for a picnic lunch, I find a Bald Eagle watching the coastal prairie from a high perch in a lone pine tree. Farther north, a strong south wind blowing across St. Marks National Wildlife Refuge probably is keeping migrants high on an overflight.

Visiting tropical birding expert Ted Parker at the Louisiana State Museum in Baton Rouge is a deliberate choice, but means missing birding opportunities along the Alabama and Mississippi coast. I am partly consoled by fair weather, for if it were foul, birding on Dauphin Island and around the Mobile delta area doubtless would have been spectacular.

Veteran birder Peter Alden commented some time ago about spring in this part of the country: "Northbound songbirds along the western Gulf of Mexico, sometimes encounter severe headwinds from the North, forcing millions down into sparse coastal vegetation along the coasts of Texas, Louisiana, and Alabama. Birders in such places as Rockport, High Island, Grand Isle, or Dauphin Island have a bonanza in both numbers and variety at such times with trees and shrubs bursting with colorful birds." That's what birders dream about.

Further consolation for missing my Alabama bonanza is a guided tour around the Louisiana museum's superb collection of bird skins, most of which are from the western hemisphere. Although John James Audubon and his cohorts shot birds in order to identify them, I hadn't really appreciated the importance of a good skin collection for ornithological research until I had seen it through Ted's eyes. There are incredible geographic variations of plumage in some species which makes it difficult to determine when a species is a species. I am privileged to look through some of John O'Neil's original plates for the forthcoming *Birds of Peru* that Ted is working on.

Next stop is New Orleans. With Charles Frank, well-known sportsman, Louisiana duck decoy expert, wildlife photographer, and artist, I have a rare opportunity to get a boat's-eye view of the Mississippi delta area. Brown Pelicans present no big deal for Floridians, but few of us have seen them nesting. Wearing waders, we haul camera equipment from boat to shore and set up a photographic beachhead for the day. Hundreds of pelicans, mostly brown but a few white, are in various stages of nest building and chick rearing. Great Egrets, in the height of their nuptial displays, share the nesting site on Queen Bess Island this spring day. The island is wall-to-wall birds: terns and gulls screaming overhead, Willets, Whimbrels, Short-billed Dowitchers, Reddish Egret, and Black-necked Stilts probing shallow offshore waters. A Clapper Rail pops out of one hole and into another secret place.

Heading westward, I ignore the questionable enticement of a restaurant advertising catfish and fresh alligator and opt to picnic at Avery Island, a private bird sanctuary owned by the McIlhenny family

Crowded conditions prevail on pelican nesting islands

and home of their famous Tabasco sauce. Carolina Chickadees chatter overhead and I wander under the Live Oak tree umbrellas over to see Great Egret rookery. Messy boardwalk nesting sites are a far cry from the isolated island I had just visited.

My route takes me along the road that hugs the coast and I watch both Glossy and White-faced Ibis as they join other long-legged waders. There is time the following day at the Sabine National Wildlife Refuge to sort out the two ibis. A slight detour could take the traveler to the Lacassine National Wildlife Refuge where air boat rides bring visitors to Roseate Spoonbill rookeries and, if it's early enough in *APRIL,* there will be geese, Fulvous, and perhaps Black-bellied Whistling-Ducks. Watch nesting marsh and water birds and migrant warblers, particularly after a cold front, at Lacassine and Sabine National Wildlife Refuges. A longer detour from the coastal route up to the Kisatchie National Forest might get you Brown-headed Nuthatch and other hard-to-find species like the Red-cockaded Woodpecker and the Bachman's Sparrow.

After a quick ferry crossing of the Calcasieu Lake outlet at Cameron, the coastal road passes a tiny woodland area at beach edge in what is known as the cheniers, unique sand dunes topped with stands of live oak trees. In the spring, these wooded areas are marvelous migrant "traps." Researchers from the University of Southern Mississippi have set up mist nets and are banding and weighing birds. A lovely mixed collection of migrants have selected the Holleyman Nature Reserve as their landing spot: Orchard Oriole, Hooded and Worm-eating Warblers,

Researcher Frank Moore extracts migrating Scarlet Tanager from mist net in anticipation of weighing and banding it

Blue-gray Gnatcatcher, Common Yellow-throat, Eastern Kingbird, Wood Thrush, Scarlet and Summer Tanagers, Painted Bunting, and Ruby-throated Hummingbird made their unscheduled stops during my hour or so visit.

The surrounding enormous marsh area in this part of coastal Louisiana produces a marvelous variety of waterbirds many of them visible in ponds along the roadway. In grim contrast to this woodland and water paradise, oil refineries loom along the coast to the west. Louisiana birding on this trip ends with a traffic-stopping view (my car is the "traffic" this Sunday afternoon) of a Long-billed Curlew in a grassy field surrounding huge oil tanks. Onward to Texas.

THE MAIN SHOW

Mid-*APRIL* to mid-*MAY* is prime time for the main migration show in most of the United States. In the south, it's earlier, in the north, later. Except for aberrant festivals in frigid springtime places such as the Platte River, the general movement of birds is north by major flyways. Birds don't always follow the traffic rules by sticking to those lanes, and birders will find ample off-flyway opportunities across the breadth of the continent.

TITANIC TEXAS

If you had just one spring in a lifetime, Texas would probably win the hottest spring birding award although there are a couple of close seconds. In addition to sheer bigness, Texas has a lot going for it as far as the birding world is concerned. It has a rich stew of habitat from mountains to coastal beaches, forests, grasslands, and near desert conditions. The 100th meridian is a reference point for the meeting of eastern and western resident species, and the Central flyway lies roughly within a couple of degrees on either side of that magic line. The icing on that cake is the proximity of tropical birds that cross the border from Mexico.

Southeast Arizona seems to be a close springtime second among the binocular bunch although it does not have the species diversity a Texas spring offers. However, virtually everyone in southeast Arizona's many birding crannies will be walking, climbing, looking, talking, eating, and even sleeping with birds on the brain. Perhaps it has an advantage of relatively short distances between birding "hot spots." Fortunately, there are excellent where-to-find guides for both Texas and Arizona that are *musts* for the birder who wants to make the most of time spent in these environs.

Perhaps tying with southeast Arizona for second place, is the other truly famous spring birding "hot spot," Point Pelee, that finger of Canadian soil that extends into Lake Erie southeast of Detroit.

Early APRIL is the time to shift your travel plans into high gear. If you're doing a Texas "birdathon" on your own, it would be best to start in the eastern part of the state and proceed slowly westward as the month progresses. (Even while speeding on the interstate, it may seem as though Texas is passing slowly.)

The Lone Star state surely is the "Lode Star" state for birders. Likely you will run into birders from all over North America along with a large handful of British twitchers who converge here during this famous birding season. Well-traveled binocular toters are likely to run into friends here as they often congregate in the same places at the same time: APRIL. Swamp buggy rides are no longer being run at Anahuac National Wildlife Refuge so seeing the Yellow and Black Rails is no longer a sure thing. Check with the headquarters for advice. Even if you miss them, you'll be delighted with the rest of the rails.

The Texas coast is marvelous for migrant shorebirds. Coastal areas and nearby flooded ricefields are often crammed full of them. Practice your identification skills; learn the differences between Greater and Lesser Yellowlegs or Short-billed and Long-billed Dowitchers. Check out the Pectoral Sandpiper throughout APRIL and pick out the White-rumped Sandpiper in MAY.

Working your way down the coast to the Rio Grande Valley you will see some of the permanent residents which make the area famous among birders: the rare Red-billed Pigeon, Altamira Oriole, Mexican Crow, Great Kiskadee (a bird you just can't miss hearing even if you can't see it), and the Greater Roadrunner. Keep your eyes open for birds just passing through: Broad-winged and Swainson's Hawks, Mississippi Kites, Nashville Warblers, or Clay-colored and Lincoln's Sparrows.

APRIL, particularly the last two weeks, brings continued excitement in Texas. Treats abound at High Island, a remote beach community on the Gulf, southeast of Houston. Migrating landbirds and bad weather spell ecstasy for wet and chilled birders. Listen to this report from Greg W. Lasley and Chuck Sexton reporting on the Texas coast in *American Birds:*

> Early on April 9, a slow-moving front passed High Island. Heavy rains on the 9th and 10th grounded 200-400 warblers of at least 23 species . . . along with many thrushes, vireos, and others (e.g., 315 Wood Thrushes and 405 Red-eyed Vireos on the 10th).

Dates for the best fallouts change from year to year, largely dependent on the weather. That report continued, "Observers variously rated the April 20–21 or the May 1–2 fallout as the best, depending on their own findings." The following year, this expectant birder guessed wrong on both weather and dates: The weather was calm and sunny, the dates APRIL 12–14, and the schedule read, "press on regardless."

A blustery northerner hitting the Gulf Coast at just the right time in APRIL can cause a "fallout" of birds of enormous proportions. The storm forces birds to drop down into small wooded areas, the first green in a while. Birding hotlines crackle and a vacant room at *the* motel in High Island is to be prized. Smith Woods and Audubon Woods (officially the Louis Smith Sanctuary), operated by the Houston Audubon Society, offer sanctuary to the birds and you will quickly follow fellow birders from around the country, and England too, as they shuttle between the two places.

If this is your first experience at a migratory "hot spot," you may wonder if these people are hunting for pots of gold as they creep stealthily along wooded paths, craning their necks to scan the canopy or crouching at the slightest scratching signal. Their treasure is the golden flash of many of the wood warblers: Prothonotary, Golden and Blue-winged Warblers, Swainson's, Worm-eating, Tennessee and Nashville (I admit to difficulty in remembering which is which), Black-throated Green, Cerulean, and Blackburnian, along with many of the rest of them. One year it was a crimson flash as *American Birds* reported 40 Scarlet Tanagers in one mulberry tree.

The trick is to listen to the weather report. Birders who have been there have been ecstatic: "There were a hundred birds on every tree at High Island when I was there last year!" *This* year, the weather may be balmy. More Orchard Orioles than you have ever seen in one place before are a bonus. You get your first full view of a Swainson's Warbler. Wood and Hermit Thrushes are about. A "Hoodie," the Hooded Warbler, is seen in some numbers, and other warblers, vireos, and flycatchers are around, but not many. You may wonder what all the excitement was about. Everyone you meet is bemoaning the fact that "spring is late this year" and are praying for a heavy storm.

Weather most definitely influences "when spring begins" for the birder. Go in late *MAY,* and the migrants will be gone and Smith's Woods silent. But in the nearby flooded rice fields there will be phalaropes and godwits waiting for that secret signal that will send them flying to the high Arctic for their breeding duties. There are always birds to watch, enjoy, and count. Imagine counting over 500 chestnut-bellied Hudsonian Godwits in one field! The peripatetic birder will see the handsome Godwits in *JUNE* at Churchill on Hudson Bay in Canada.

North of Beaumont, not very far north of High Island, is the Big Thicket country where some very special birds can be discovered. The Red-cockaded Woodpecker, Brown-headed Nuthatch, and Bachman's Sparrow are very special for most birders. Don't expect them to be sitting on every tree branch. These rare and elusive species are worth some homework on habit and habitat as well as song. It may take some time and work to get them.

If you are not headed directly west, try working your way down the Texas coast to Brownsville and the Lower Rio Grande' Valley. In late *APRIL* 1988, Lasley and Sexton reported a formidable flock of 700 to 800 Mississippi Kites at Bentson State Park, one of the hottest Texas birding spots.

Migrating wood warblers, tanagers, orioles, buntings, and flycatchers are common in many parts of Texas, most of them fairly widespread. For the migrants you are likely to encounter on your spring drive through Texas, check particularly the Lane guides for the month and place you will be traveling. Often other, very localized guides and checklists are available to help you. Also check your field guide range maps for those blank areas south of the summer breeding range; that is where some of the migrants will drop down to get out of the weather or just to rest and feed.

For the birder filling in a life list with the rare and hard-to-find species, try to tie in to your spring Texas trip a stop at the Attwater Prairie Chicken National Wildlife Refuge west of Houston. Because the population of the Attwater race of the Greater Prairie Chicken, once so common on the Texas grasslands, has declined precipitously, it might be better to go with a birding group to see this fine grouse. Most of the refuge has been closed during the breeding season so if you try to do it on your own, be sure to call in advance. Otherwise, you may find that you're just driving through the grasslands and seeing nary a chicken. One birder of exceptionally close acquaintance didn't. She tried desperately to turn some Bobwhites into prairie chickens and missed Le Conte's Sparrow and White-tailed Hawk. She ended up being delighted with Sprague's Pipit and Upland Sandpiper, life birds for this easterner. Lekking time lasts from *FEBRUARY* through *APRIL*, so if you don't get there in the spring, try late winter.

The Edwards Plateau (known as the Hill Country) west of the Austin/San Antonio area is another super birding area. Watch Cassin's Sparrows do their skylarking bit over a grassy field. See your first Grasshopper Sparrow singing from the fence while you photograph the wildflowers. Patriotic roadways emblazoned with red, white, and blue wildflowers encourage stopping to bird. Credit is due for much of that beauty to "Ladybird" Johnson who campaigned hard and long. (In fact, there will be no doubt in the traveler's mind that this is "Johnson country.") As everywhere, both permanent and migratory birds are easier to see in the springtime when they are full of song.

Late *APRIL* through early *MAY*, if you are following the Texas interstates and byways westward, will find you in the "real West," terrain familiar from all those western movies. Western migrants will be peaking in Big Bend National Park the first week in *MAY*. Try to be there within a week or so on either side of that week for the best spring birding.

For keen watchers, it is the Colima Warbler that is the prime object of that trip down through the marvelous desert country dominated by the Chisos Mountains. Late *APRIL* through *MAY* is the time to try your hiking legs on the arduous trail up to Boot Spring Canyon where the warbler will be singing. How many of today's birders bemoan their years ago visit to Big Bend National Park before they achieved the birding sophistication needed to see this drab warbler in its only known breeding spot north of the Rio Grande?

Big Bend holds the national parks record for the top number of species. "Good birds" found in Big Bend in the springtime include the Lucifer Hummingbird, Gray Vireo, Varied Bunting, Crissal Thrasher, Elf Owl, Western Screech-Owl, Vermillion Flycatcher, Pyrrhuloxia, and Rufous-crowned Sparrow. Don't forget that there is only a few feet of muddy water separating the Mexican birds from what we consider the North American birds, so take your Mexican field guide for extra verification. Look out for the Aztec Thrush, Rufous-capped Warbler, Rufous-backed Robin, and Black-vented Oriole.

HOT TIME IN ARIZONA

Late *APRIL* through *MAY* is one of the better times to aim for southeast Arizona although that lovely part of North America has jewels to offer at almost any time of year. It's hot in most of Arizona during the spring and hotter in the summer. But the mountains of southeast Arizona are delightfully cool, actually cooler in summer than in spring. They offer superb birding and one visit there will convince the birder that this is a "same time, same place next year" destination. Tucson is the gateway to birding paradise in the southeast corner of Arizona.

If you're driving from Big Bend at the end of your Texas sojourn, head for El Paso on Interstate 10, with a stop in the Davis Mountains to check the Montezuma Quail and Common Black Hawk off your life list. New Mexico is still to be crossed before you reach Portal, Arizona, your primary destination. Depending on how much time you have, you can zip along or enjoy some side trips. The Pettingill bird finding guide to the western United States in hand, you might just want to make a detour to see the famous tourist attraction, Carlsbad Caverns with its stalactites, stalagmites, and Cave Swallows. Of course if you saw the swallows when you were up on the Edwards Plateau, feel free to bypass the tourists. Tiring of the interstate, loop off and visit the Silver City area with its wealth of migrants. Myra McCormick, long-time birder who lives there, pinpoints the arrival of early migrants: *MARCH* 8 for the Broad-tailed Hummingbirds near Pinos Altos, and *MARCH* 25 for the Turkey Vultures and Painted Redstarts in Glenwood. Listen in the evening at McCormick's Bear Mountain Guest Ranch for the Buff-collared Nightjar.

At Portal, gateway to the Chiricahua Mountains, you will quickly discover hummingbird heaven from spring through late summer. These nectar sippers will be visiting feeders right outside your cabin window. Campers in Cave Creek Canyon will find their hummingbirds nearby. Several birder friendly residents in the little town welcome visitors to their sugarwater feeders, and the Southeast Arizona Research Station, a short distance away, provides endless photographic opportunities.

Southeast Arizona is a fine place for many of the Mexican species which can be found only here. Don't worry about finding the Elegant Trogon. At the Portal General Store, the standard greeting is, "Have you seen the trogon?" If you haven't, full information is forthcoming on where it's being seen today. Entering Cave Creek Canyon late in the afternoon, its flaming rhyolite cliffs dazzle the eye, making the dark forest of the campground seem more mysterious. Friendly birders quickly dispel the mystery of finding the trogon and will volunteer up-to-the-minute directions.

A sip of sweetness and thou watching from the window

Walking up the path, watch the flashy Painted Redstart flitting along the creek ahead of you. Just after the path crosses the creek, you approach a quiet group sitting on the stream-side boulders holding their glasses expectantly. "The nest hole is right up there. It's been about 45 minutes since one left and there should be another changing of the guard in about 5 minutes." You settle as comfortably as you can between a rock and a hard place with a straight shot at the hole. Soon the trogon swoops in and, following a momentary greeting, its partner immediately flies off presumably to seek the next course. Trogons arrive here towards the middle of *MAY*.

If you were there in *MAY* 1985, you would have shared the excitement of seeing the Flame-colored Tanager making its first recorded foray across the border. Well-known birder, Robert J. Morse, who lives there, identified the visitor from Mexico and the word was out. It's nice to have a cooperative rarity, and this one provided superb views for the hundreds who traveled to see it. A climb up the mountain at Rustler Park with wildlife artist Lydia Thompson and her father/birder John, produced views of Pygmy Nuthatch and the visiting Mexican Chickadee, found nowhere else north of the Mexican border. A fleeting glimpse of a Hermit Warbler puts it on the "bvd" list (better view desired). Highlight of the pilgrimage the next year was the stunning Red-faced Warbler posturing before a group of initiates for at least 15 minutes.

APRIL, and as early as *MARCH,* provides fine birding in southeast Arizona, but *MAY* begins the best season when all the nesting species will be present. Owls are most easily seen now for they will be calling. Other places vie with Portal for birding fame. Each birder, and some bird species, has a favorite location: Madera Canyon, Ramsey Canyon, Patagonia, Guadalupe Canyon, and so on. It's definitely not an area you can do in a weekend. A few of the other special species you will be looking for are Hermit and MacGillivray's Warblers and Williamson's Sapsuckers.

This part of Arizona has many year-round specialties, but, as usual, they are often easier to see in the spring than at other times. Serious birders will be looking for Buff-collared Nightjar, Thick-billed Kingbird, Gray, Dusky, and Hammond's Flycatchers, Greater Pewee (if you don't see it, you will hear the spaced out "José Maria"), Northern Beardless-Tyrannulet, Bendire's Thrasher, Black-capped Gnatcatcher (be careful not to confuse it with the Black-tailed), Abert's Towhee, Rufous-winged, Five-striped (try Sycamore and Chino Canyons), and Botteri's Sparrows, lots of Yellow-eyed Juncos, and exciting hummingbirds. *MAY* is best for owls for they will be calling. This corner of Arizona is known for the "lady" warblers. Not a sexist remark at all, but Virginia's, Lucy's, and Grace's Warblers. The scientific names are unmistakable: *virginiae, luciae,* and *graciae*: a daughter, a wife, and a sister.

PELTING PELEE

MAY is the critical time to visit another famous spring birding "hot spot" a long way from Arizona. The first two weeks are generally considered by many to be the best at Point Pelee, Ontario, Canada's southernmost point of land. Experts give the edge to the second week but each day brings in different species so plan to be there a minimum of one week, preferably two. Peter J. Hamel reports in the February 1989 issue of *Winging It,* newsletter of the American Birding Association, that he clocked 208 species in *MAY* 1988. If you can't make it there in prime time, any time between *MARCH* and early *JUNE* will provide you with a prime birding experience.

Migrant birds winging north across Lake Erie, touch down in astounding numbers on this finger of land jutting into the lake from the northwestern shore. It is an easy drive southeast from Detroit or southwest from Toronto. Surrounded by woodlands and meadows, the largest habitat is the vast freshwater marsh that welcomes the tired travelers. Roughly 350 species have been reported here including a superb collection of wood warblers.

Pelee is where many British twitchers choose to take their first North American birding trip. Canadians and those birders who live in the northern states swear by this remarkable place. "For the visitor to North America on a first trip from abroad, and not knowing anything about the birds of that continent, Point Pelee is probably the best place to start. They can see a wide variety of families and species in a very pleasant place and usually with plenty of helpful American or Canadian birders to give advice." So writes Peter Carlton, British birder who has visited Pelee three times. When birders speak of Pelee, they're not just talking about Point Pelee National Park but about many places in the area. Veteran birding duo Mary and Tom Wood favor the onion fields north of the park, the Hillman Marsh Conservation area, and Kopegaron Woods. Late *APRIL*/early *MAY* will catch the last of the waterfowl passage and many raptors will be transiting overhead. In early *MAY*, as many as 35 species of wood warbler along with tanagers, buntings, and finches will be dropping in for a visit. Later in the spring, you can sort out the pesky (at least early birders think so) *Empidonax* flycatchers, most easily identified by their distinctive songs.

Looking for specific species? Check a guide book to ascertain whether it is best to be there early or late *MAY*. For example, Blackburnian, Connecticut, and Mourning Warblers will meet you at Pelee in the latter part of the month while the Nashville Warbler will be there a bit earlier. Given a couple of weeks, you will see species nearly matching in number the days of the year. Many, such as the Henslow's Sparrows, often stay to nest and no matter when in the spring you are there, you will likely encounter stragglers of many species.

As is often the case when migratory birds are crossing bodies of water, bad weather is good news for birders. A northerly storm tires the birds battling against it and they will be forced down on the point. You may plan a few days or a few weeks at Pelee. The longer you stay, the more you will be able to take advantage of the weather rather than the reverse. Appreciate the different waves of birds coming in, each species with its own timetable, a wondrous occurrence.

Mid-*MAY* onward is Kirtland's Warbler time in Michigan's jackpines in Huron National Forest, easily reached from Pelee. The rare and endangered Kirtland's Warbler is relatively easy to see and easier to hear, so best to be familiar with its song. Don't however, play a tape while you're driving along Stephan Bridge road east of Grayling. Michigan birder Bill Bouton points out the irony of allowing "tanks, trucks, aircraft, and bombing by the national guard" but not bird tapes. I guess warblers don't respond to bangs and booms. If you don't want to chance it on your own, contact the Michigan Department of Natural Resources

(P.O. Box 507, Grayling, MI 49738) for information on their free tours which operate through *JULY* 4.

Continue north across the Mackinac bridge to another superb birding spot, Whitefish Point Bird Observatory. Jutting into Lake Superior in Michigan's Upper Peninsula, it is a meeting point for southern and western species, and for Arctic and boreal species. As Pelee is a landing point for northbound migrants, Whitefish is a take-off point. Whitefish also has a spring hawk watch from *MARCH* 15 through *JUNE* 1. There will be rousing raptor action, but remember that each species has its own schedule and flights will peak at different times. Sharp-shinned and Broad-winged Hawk are the most commonly seen. However, some special out-of-range rarities have been Swainson's Hawk, Gyrfalcon, and Prairie Falcon. Locals have welcomed other out-rangers such as Townsend's Solitaire, Rock Wren, Clark's Nutcracker, White-winged Dove, Sage Thrasher, and Sprague's Pipit. David J. Powell reported in *American Birds* (fall 1987) the probable sighting there in *MAY* 1987 of Pacific and Yellow-billed Loons. Definite sightings were 13 Boreal Owls between mid-*APRIL* and early *MAY.*

APRIL, according to Bouton, is a good time for most eastern and northern owls, lots of waterfowl, and the handsome Red-necked Grebes. *APRIL* is fine for northbound raptors that continue through until the end of *MAY.* By this time breeding species such as Le Conte's Sparrows, Yellow-bellied and Olive-sided Flycatchers, Merlins, and warblers have established territories. Depending on the weatherman, migrant passerines are beginning to come through, peaking between *MAY* 12 and *MAY* 20.

At the same time of year, another nearby birding treat is Long Point, Ontario, at the northeastern end of Lake Erie. Canadians visiting Sanibel Island, Florida, for some nice warm winter birding, highly recommend Long Point. Recommendations from locals are to be prized. Canvasbacks and Tundra Swans use Long Point as a major staging area for flights north. In Pennsylvania, on the southern shore of that great lake, the Erie National Wildlife Refuge offers another side trip. You will see songbirds and perhaps the rare Henslow's Sparrow. Try also the Presque Isle State Park on that little hook of land jutting into Lake Erie near the lake's namesake city.

SIDESHOWS

These three famous birding "hot spots" are not the only places to be in springtime, nor is the magic month of May the only "best" time.

Throughout the season, all across the country, binocular grabbers peer skyward, into blooming bushes, up and down coastal beaches, across lakes and rivers, and across the waves on pelagic trips.

SPRING HAWKING

Early *MARCH,* with the snow not yet off the ground in many places, well into mid-*MAY,* depending on where you are, is generally considered to be the time to scan the skies. Eagles, Northern Harriers, Cooper's, Sharp-shinned, Red-tailed, and Rough-legged Hawks, Osprey, American Kestrel, and Merlin can be found pretty much all over the country. Broad-wings will be more common in the east, Swainson's and Ferruginous Hawks and Prairie Falcons in the west. Peregrine Falcon sightings anywhere are cause for excitement. Some hawks just don't migrate, others are local in their distribution. Hawk watching is more spectacular in the fall, but eagle eyed birders set up hawk watches in the spring too, often reporting sizable numbers.

Watching eagles, hawks, falcons, and even vultures wafting across an azure sky offers one of the universal thrills of birding. For the early birder, it is the sight of a raptor overhead that can be positively rapturous. For the experienced birder, the sight is no less spectacular, but the challenge of identification has led to something of a hawk watching cult. So enthused are hawk watchers about their biennial ventures that they formed an organization, the Hawk Migration Association of North America, (HMANA) You can join it, peruse their publications, study their slides, and show up at regularly established hawk watches. Hawk watching tends to be more spectacular in the fall, but spring offers plenty of opportunity across the country to watch these kings of the sky.

Where to see them in the spring? Aside from brief references throughout this chapter, check information in *American Birds* and in HMANA publications for the best spots and the peak times for the species you particularly want to see. Hawkwatchers not only want to identify a hawk they haven't seen before, they go bonkers over numbers. You would too if you had been watching over 3,000 Broad-winged Hawks fly over Bentsen State Park in Texas on April 8, 1987, or if you had clocked over 2,000 Sharp-shinned Hawks between nine and ten A.M. from April 8 to May 15 at Whitefish Point in Michigan, as reported by HMANA.

OFFSHORE

If winter seems to be lingering longer than usual where you live, early spring offshore of the continental United States may be appealing.

Miami is a jumping off point for birding areas which may provide you with your first taste of "foreign" birding.

MARCH is sort of an inbetween month. It's beginning to be a migrating month in southern states but migration hasn't picked up much speed. Wintry cold still blankets much of the northern two-thirds of the continent. For North Americans, it may be just the right time to visit some of the nearby "foreign" places in between the western hemisphere continents.

If you didn't get down to Mexico and Central America last winter, plan a trip to Costa Rica in early spring. This is a favored time in a friendly country. From marshes to mountains and from rain forest to cloud forest, this tiny country hosts a tremendous variety of bird species (over 800). With good guidance, you should be able to see more than half in a couple of weeks. Seeing the Resplendent Quetzal, even if it's just an iridescent green image slowly passing distant dark green mountains, is a thrill every birder anticipates.

You know what Scarlet Macaws look like, and you will probably recognize one instantly. One can guess what a Black-chested Hawk might look like if you saw one, but how about a Double-striped Thick-knee? Watch intently for a turkeylike Black Guan scuttling through the underbrush, and listen for a Three-wattled Bellbird singing its ringing metallic song from the top of a tall laurel tree. These are just the beginning of Costa Rica's store of jewels.

APRIL is a lovely month in Jamaica and Puerto Rico, offshore island-countries favored by those who want a taste of foreign birding but don't want to stray too far from home. Both native and migrant birds are there and the weather is not too hot. James Bond, namesake inspiration for Agent 007, wrote a fine guidebook for these and other Caribbean countries. Jamaica has 25 endemic species but to see them all, crisscross the island from end to end and visit the beaches and the mountains. You can figure out many of the endemics just by the names: Jamaican Todies, Orioles, Vireos, Lizard Cuckoos, Owl, Crow, and Blackbird—all with Jamaican as a first name. You may not think a blackbird would be a particularly interesting bird, but as the early morning mist rises from the mountains in the Cockpit country, the Jamaican Blackbird flies across the valley uttering a peculiar rattling cry that is worth the price of the trip just to hear it. Watch also for the Jamaican Mango, a hummingbird not a fruit tree. The Streamertail, another hummingbird, is Jamaica's national bird recognizable instantly when you see it. Puerto Rico has its share of endemics to which its name is attached—Puerto Rican Lizard Cuckoo, Screech-Owl, Tody, and Tanager. Sophisticated spotters will be particularly looking for the endangered Puerto Rican Parrot.

Colorado Grousing

When it's springtime in the Colorado Rockies—and out on the plains—birding enthusiasts peer through an occasional late snow storm to watch the last of the wintering species, such as the Rosy Finches, depart for points north, and to watch migrating waterbirds, shorebirds, and songbirds coming up from the south.

Large concentrations of waterbirds arrive as early as *MARCH* in Colorado although many have wintered over. In *APRIL,* Clark's and Western Grebes can often be seen together on reservoirs and ponds. Scope the pond carefully to identify the Clark's. Then watch in amazement as those black-and-white creatures rise up to dance on the water's surface in the water ballet of a mating display. Wilson's Phalaropes will be spinning in the spring in shallow ponds easily seen from many back roads. There were 2,000 in Lamar on May 18, 1986, according to Hugh E. Kingery reporting in *American Birds.*

APRIL is a fine time to be in Colorado, a big state more famous for its ski slopes and summer camping than for its bird life. Such lopsided fame is really unfortunate, for Colorado trails only California, Texas, and Arizona in numbers of bird species. Watching spring unfold in Colorado is a lovely experience. As the streams begin to flow again, the leaves on the bare trees almost unfold and open up before your eyes.

APRIL is particularly fine for the grouse crowd although these chickenlike birds "boom" at their leks from *MARCH* through *MAY.* One of nature's awesome spectacles is a Sage Grouse lek. A lek is where male grouse and other chickenlike birds gather in astounding numbers in the chill early dawn to strut about booming a low message to each other and for the benefit of any females chancing to be around. "I can puff up my breast bigger than you can," they must be saying. Watching the courtship puffery from your vehicle as the light gradually bathes the landscape is to experience what might almost be described as ancient druid ritual hidden from the rest of the world. If you are not with a birding group, inquire of local experts about the location of the leks as they are often difficult to find in the dark.

"Chicken-chasing" is a spring specialty in many parts of the west and has considerable variety in Colorado. Lesser Prairie Chickens boom in the Comanche National Grasslands in southeast Colorado, the Greater Prairie Chicken prance near the northeast corner of the state, and Sharptailed and Sage Grouse leks can be found on the eastern slope in what is known as North Park.

Don't make the mistake of thinking that Colorado is all mountains. Vast plains, grasslands of old, cover the eastern third of the state. They

Sage Grouse puffery on an early morning lek

provide marvelous birding year-round, but especially in *APRIL* and *MAY*. If your bird list is short on longspurs, you can pick up Chestnut-collared and McCown's Longspurs in Pawnee National Grasslands. Best to catch them in spring when they are singing. Scan the prairie carefully for a Mountain Plover peering, motionless, over the grass.

APRIL in Idaho is for birders looking for sterner stuff and some springtime snow. Birds in Idaho? You bet. Birding to give you thrills along with the chills. A whopping two-thirds of the state's land is under federal management and no matter where you travel, there will be a wildlife area nearby. The Snake River Birds of Prey Area near Boise is the nesting site of one of the world's densest populations of birds of prey: between 600 and 800 pair. Mid-*APRIL* brings large concentrations when combined with fairly decent weather, but any time from mid-*MARCH* until early *JUNE* can be productive.

Southcentral Idaho in the Twin Falls/American Falls area provides excellent birding. Bald and Golden Eagles, along with most of the raptors found in the west, will be circling overhead when the sun comes up. On the lakes and waterways, graceful Tundra and Trumpeter Swans along with Horned, Eared, Clark's, and Western Grebes, in addition to huge concentrations of ducks, will be performing their courtship displays. Barrow's Goldeneye would warm your binoculars.

At the Minidoka National Wildlife Refuge alone, some two hundred thousand ducks and geese can be seen during spring migration.

Spring nights in Idaho hold the promise of owls and you may add sought after Great Gray and Boreal Owls to your list. Long-eared and Short-eared, Northern Saw-whet and Western Screech-Owl are possibilities. If you're looking for a Chukar or Gray Partridge, try the Bear Lake National Wildlife Refuge; it's a possibility but not a certainty. Sandhill Cranes will be courting and there may even be a Whooping Crane among them.

FORTING AROUND

APRIL quickly merges into *MAY* and ardent birders face tough choices on where to take their binoculars. One of the favorite, though more remote birding "hot spots," is the Dry Tortugas, tiny islands at the far tip of the string of islands known as the Florida Keys. Fort Jefferson, a fine example of a 19th Century fortress will interest history buffs, but it is noddies and boobies and tropicbirds that draw bunches of birders. Most will take one of several boats from Key West, boarding at midnight and arriving at Fort Jefferson at dawn. Bunks on the small boats are less than luxurious and a sleeping bag may earn you a place on the

Fort Jefferson dominates Dry Tortugas, where migrating birds rest and pelagic birds nest

airy deck. It's possible to fly out to the Tortugas, but if you plan to stay more than a few hours, take your sleeping bag, drinking water, and tent, for there is nothing vaguely approaching a Tortugas Hilton.

Migrant land birds often stop over in the Tortugas, but it is the pelagic species that catch sharp eyes. A huge Sooty Tern colony on Bird Key is impressive. Brown Noddies are common and if you study the mangroves very carefully you might spot the rare Black Noddy. Within an hour of arriving in the Tortugas in 1985, our Bird Observer group saw three White-tailed Tropicbirds. You couldn't miss them as they screamed their way over the fort. Had we been there a day later, we would have missed them. Brown Boobies perch on old pilings around the fort while Masked Boobies roost on sandy keys such as Hospital Key. The trip across Gulf waters on the return to Key West may well produce Bridled Terns, Audubon's Shearwaters, or jaegers. On the 1989 "pilgrimage" to Fort Jefferson, several Northern Gannets delighted a Four Points tour group on the return trip. Numerous local and national birding groups schedule trips to the Tortugas. Such groups help cope with the lack of food, water, and accommodations, but if you're a good camper, tenting under the stars outside the walls of the fort on Garden Key can be a more aesthetic experience than sleeping in the boat's crowded bunk room.

APRIL and *MAY* offer delightful weather in south Florida so if this trip is your first visit to south Florida, do spend some extra time in this subtropical area after your Tortugas trip. Migrant songbirds will be passing up the peninsula. Watch in *MAY* for the Connecticut Warbler. Mop up Florida's spring and summer specialties: Gray Kingbird, Black-whiskered Vireo, Mangrove Cuckoo, Antillean Nighthawk, and Swallow-tailed Kite. Useful locations include the Florida Keys, Everglades National Park (even though it's getting a bit hot and buggy by this time of year), Key Biscayne across a causeway from Miami, Loxahatchee National Wildlife Refuge where Fulvous Whistling-Ducks are getting ready to vanish from accessible areas, and Sanibel Island where the Roseate Spoonbills will be showing. Check your Lane guide for directions.

Like most tongues of land, south Florida gets its share of offshore species from time to time. In the springs of 1988 and 1989, it was two species of swallows swirling around noisy turnpike traffic that stimulated birders traveling glands. Cave Swallows, primarily a Central American and Caribbean species, (this one soon to be split from the Cave Swallows seen in the west) had been nesting in recent years under the Florida Turnpike near Homestead. Although they are enough to attract attention, it was the vagrant Bahama Swallow, looking like a cross between a Tree and a Barn Swallow, that provided the double bill to tantalize tickers from all over the country.

Fulvous Whistling-Ducks rest in Loxahatchee National Wildlife Refuge

UP EAST

APRIL is a fine time of year for a pelagic trip off the Maryland and Virginia coasts. Anyone who predicts "good" sightings on any given pelagic trip should be prepared to be dumped into Davy Jones locker, but records from previous years suggests optimism for Sooty Shearwaters and Wilson's Storm-Petrels. Red Phalaropes are also a possibility and Northern Gannet should be easy. If you're interested in spring pelagic birding on the east coast, contact one of the tour operators listed in Appendix 2 at the end of this book, or contact Audubon Societies or bird clubs in coastal cities, some do offer ocean trips.

Wildlife refuges are located all along the Atlantic coast. The Savannah Coastal Refuges in early spring attract Black and Surf Scoters, Hooded and Red-breasted Mergansers, Ruddy Ducks, plus many shorebirds, gulls, and terns. Dip in and out of the refuges, some of which may be reached only by boat. Many eastern woodland species including the Brown-headed Nuthatch will be found along this coast in the spring. On a drive north, try visiting as many refuges as time permits.

Charleston, South Carolina, has long been a favorite with birding and history buffs. The city springs to life with colorful blossoms on nearly every tree and bush. Marshes at nearby Cape Romaine National Wildlife Refuge are also full of life. Longlegged waders stand like sentinels in their watery world that stretches as far as the eye can see. Wood Storks, Oystercatchers, and Marbled Godwits are there in considerable numbers. Brown Pelicans will be conspicuous. Black Skimmers skim the water surface for supper.

Mid-*APRIL* to mid-*MAY* in Virginia is the time to visit the Great Dismal Swamp National Wildlife Refuge for an assemblage of eastern migrants. Some stick around for short stretches, others stay to nest. If you're persistent, you would find 17 species of nesting warblers and vireos including the Prothonotary, Swainson's and Worm-eating Warblers. The American Woodcock reputedly nests there too.

MAY is the month for miniature, rather than massive, birds in Delaware Bay, particularly at Reed's Beach in New Jersey north of Cape May. Around the third week (aim for the full moon) in the month is the peak period to witness a stunning gathering of shorebirds in eastern North America. An estimated half million (some say a million) gather here at high tides to fatten up on horseshoe crab eggs before heading to their arctic breeding grounds. During the last couple of weeks in *MAY,* you will see Sanderlings, Semipalmated Sandpipers, Red Knots, Ruddy Turnstones, and Dunlins, all just up from their long flight from South America. Here they gorge themselves before flying off on the last leg of the trip to their arctic breeding grounds. Roger Tory Peterson described the beach as "a magnet in spring to the greatest concentrations of migrating shorebirds in eastern North America." While in the neighborhood, don't overlook the Cape May area for migrating warblers. Although best know as a fall hawk watching site, there is a fine spring movement of songbirds through this southern tip of New Jersey.

Any chance to be in the New York metropolitan area in the spring (and again in the fall) is reason for both urban and suburban birding. Try Central Park in *MAY* and watch businessmen carrying brief cases and binoculars spotting spring warblers. Out in Jersey, try the Great Swamp National Wildlife Refuge near Basking Ridge in *MAY* for a fine collection of warblers and other songbirds. Off Long Island shores, waterfowl will be gathering up their pluck, and whatever else they can find to motivate them, and heading for northern nesting sites.

MARCH to mid-*MAY* in the northeast coastal area provides surefire shorebirds. Just pick your spot: Parker River National Wildlife Refuge on Plum Island north of Boston is a favorite. Watch migrating waterfowl, raptors, and in *MAY,* be there for the warbler migration.

Late *APRIL* to mid-*MAY* is the time to be up at the Moosehorn National Wildlife Refuge on the easternmost point of Maine. The main show there is the spectacular soaring courtship dance of the American Woodcocks. Rangers say you're guaranteed to see them. Be there before dawn or just after sundown. Moosehorn is the only national wildlife refuge where the woodcock is king. While you're there, wander through the refuge for encounters with many of the eastern woodland migrants. Heading back down the Maine coast, take a trip out to Monhegan Island east of Portland in late *MAY* for woodland species.

MIDSECTION MIGRATION

All along the Mississippi flyway in the spring, waterfowl and shorebirds move steadily northward. Review your national wildlife refuge reference books and if you're planning a trip through this part of the country, send for information from the wildlife areas you will be near. Check fall issues of *American Birds* for dates special birds were seen during preceding springs. Last year offers a prediction, not a guarantee. For where migrating birds will touch down depends, in large measure, on the weather.

At the upper end of the flyway, South Dakota, known to vacationing America as a popular summer destination, the migration seasons bring birding visitors. The western part of the state is famous for the Black Hills, Badlands, Mt. Rushmore, and Custer State Park where "the antelope and buffalo" (in reality, pranghorns and bison) play. Although such are the major attractions, do take your binoculars for birds play there too.

For the greatest variety of birds on your visit to South Dakota, spend time in the eastern wetlands. During spring and fall migration, you will witness a fine movement of water species including American White Pelicans, Wilson's Phalaropes, warblers, and other western songbirds. At the Madison Water Management District, watch also for Black-billed Cuckoos, Wilson's Warblers, and Least Flycatchers. Waubay National Wildlife Refuge takes its name from a Sioux word meaning "nesting place for birds." From spring until fall you can find waterfowl, sparrows, and, during migration, many western passerines.

The heart of the Midwest—northern Missouri, Kansas, and Iowa— is known more for bird seed than for birds, but if you're driving through in the spring, migrant waterfowl and warblers are flying through in fairly large numbers. Not much place to rest in fields planted with corn, wheat, alfalfa, and sundry other seed crops, so search for suitable habitat: wildlife refuges or an occasional state park. Try the Union Slough National Wildlife Refuge up in northwestern Iowa. American White Pelicans and gaggles of geese visit each season.

The Flint Hills National Wildlife Refuge, Kansas, is not a particularly beautiful place but its rather scruffy habitat is attractive to migrant birds and its location near several of the larger midwestern cities makes it attractive to birders. Kansas is a member of the "400 club" of states so a lot of migrant birds will be passing over those cash crops and livestock both spring and fall. Migrant passerines passing through in *APRIL* and *MAY* will be in satisfying, if not spectacular, numbers. Get in the right marshy area at this time of year and you may see Dickcissels crowning every high reed in sight—as many as two hundred of them, singing their distinctive song that sounds roughly equivalent to its name. Related to grosbeaks and cardinals, these small brown birds with yellow breasts nest here as they do in a broad area of the central United States.

The midsection of the continent offers many opportunities to practice identifying look-alikes. Fairly obvious to the reader of North American field guides is the fact that some species are found in the east and others strictly in the west. Field Sparrows are primarily eastern species, Brewer's Sparrows are western birds. For a reference point, get out your atlas and check the 100th meridian, the theoretic division between eastern and western species. Eastern Meadowlarks and Western Meadowlarks are two separate species. The two meadowlarks look pretty much alike but their calls are quite different so check out your bird call tapes. The Baltimore Oriole of the east and Bullock's Oriole of the west were lumped a few years ago as the Northern Oriole. Recalcitrants like to note and distinguish between the two races of this and other species, hoping that someday they will be unlumped. Same thing is true with Myrtle and Audubon's Warblers, formerly separate races, now known collectively as Yellow-rumped Warblers. Try the Swan Lake National Wildlife Refuge in north central Missouri for good views.

As the Mississippi River is both lure and guideway for northbound species, so it is that the Great Lakes lure avian migrators. Visit areas south of the lakes and along southern lakeshores. Stop by Crane Creek State Park southeast of Toledo, Ohio, for eastern migrants. Hawks, going through in early *MARCH*, can be seen in goodly numbers at places like Dunes State Park, Indiana, at the southern shore of Lake Michigan.

Tundra swans gather in large numbers in mid-*MARCH* at the eastern end of Lake Erie and to the east along the Susquehanna River near Lancaster, Pennsylvania. Tundra Swans are known to get together there in numbers of up to ten thousand before heading for their nesting sites along the northern borders of Canada and Alaska. It's a mighty impressive sight.

From mid-*MARCH* to mid-*APRIL* is the time for another spectacular in some unlikely spots. Greater Prairie Chickens are on their booming grounds at the Ralph E. Yeatter Prairie Chicken Sanctuary in southeast Illinois. Viewing is permitted but limited. You will need to make reservations in January or February by contacting the Illinois Natural History Survey, 304 Poplar Drive, Effingham, IL 62401, (618)783-4125. At the same time, these grouse are also booming in nearby Missouri at the Taberville Prairie Chicken Refuge near Clinton. Late *MARCH* through early *APRIL* is the time to see Sandhill Cranes in places more accessible to easterners than is Nebraska's Platte River: the Jasper-Pulaski State Fish and Wildlife Area. Southwest of South Bend, Indiana, this area is good for not only cranes but the accompanying waterfowl and marsh birds.

Spring at the Horicon National Wildlife Refuge in Wisconsin follows a pretty typical spring pattern. Ducks begin dribbling in early in *MARCH*. If your passion is redheads, Horicon is the largest nesting area for Redheads, a handsome duck with a rusty-red head. Huge numbers of Canada Geese pass through. Hawks, a few shorebirds, and early warblers fly in during *APRIL*, but the main migrating warblers, vireos, and other songbirds turn up in *MAY*. Florida birder Randall M. Evanson, who leads WoodStar Tours, gives some insider advice on Wisconsin birding: "We hope that you will not overlook one of birding's best-kept secret 'hot spots', the northern Door County peninsula of Wisconsin [it's the 'thumb' sticking up into Lake Michigan]. During migrations one may see everything from yellow rail and least bittern and white-winged scoter to Kirtland's and hooded warblers and upland sandpiper. Several times I have tallied more than 80 species on a week-end." Wisconsin is a vacation state and many birds like it there so well that they stay to nest. It should be easy to watch Forster's Tern, Black-billed Cuckoo, lots of flycatchers including the Willow, both meadowlarks, Rose-breasted Grosbeak, and Indigo Bunting.

Millions of ducks stream northward through Ontario and Manitoba to the eons-old potholes and marshes that make the northern prairies the "duck factories" of the continent. Winnipeg is a convenient staging area. Try Oak Hammock Marsh, a short distance north of the city, for Black Terns, raptors, and Yellow Warblers singing in the willows. *American Birds* reported large concentrations of Rough-legged Hawks there in mid-*APRIL*, 1988. In late *MAY* to early *JUNE*, invest a few days in the area on your way up to see the Ross' Gulls in Churchill. You will be going through these parts anyway to catch plane or train to that far-northern summer "hot spot."

PACIFIC COASTING

APRIL and *MAY* in California, as is the case in so much of the country, is one of the best times of the year although some spring migrants begin singing their way north in early *MARCH*. Resident species are also singing if they can carry a tune and some winter species are still hanging around. California is a most flexible state: it will adjust to your timetable. This is just another way of saying that whether you're on the coast, in the mountains, or desert, or whether you're there in spring, summer, winter, or fall, California will statisfy. Like spring Atlantic coastal birding, spring Pacific coastal birding fairly bubbles with birding activity.

The Salton Sea, a saltwater lake way down at the southern end of the state, is one of the most famous of the migrant "traps." The lake traps millions of birds flying through the area during migratory seasons —swallows, shorebirds, songbirds. Watch for Vaux's Swift in late *APRIL* and early *MAY.* That is also the time to watch and listen for Warbling Vireos, Nashville, Townsend's, MacGillivray's, Wilson's and other warblers, and Western Tanagers. Some oddball birds to be found this far inland include Pacific Loon and Pomarine and Parasitic Jaegers.

Coastal migrants may be heading all the way north to Alaska. For those oceangoing migrants, take a pelagic trip. Shorebirds will be headed from their wintering areas in Central and South America for Gray's Harbor on the Washington coast, a major staging area on the very long route to high Arctic breeding areas. Offshore life listers begin in *MARCH* to look for a seagoing auk, Xantus' Murrelet. Xantus was a Hungarian who collected birds in the California area in the 1800s and for his efforts, is forever remembered for the murrelet and for a hummingbird. Lucky birders will see the Xantus' Murrelet bobbing on the water until *JULY.* Sooty Shearwaters are often seen in large numbers anywhere along the California coast. Several trips go out from Newport, Oregon, and Westport, Washington. Add the Black-footed Albatross to your list.

The northwest coastal area has a large number of resident bird species (see Chapter 7), but the spring brings some specialities. During the first half of *APRIL,* visit Vancouver Island for large numbers of Pacific Loons. Later in the month, search coastal areas in Oregon for the rare Yellow-billed Loon. In early *MAY,* huge congregations of shorebirds dominate the Columbia River basin dividing Washington and Oregon. Jaegers may come close enough to shore to be seen by the earth bound birder.

ALASKA SUPERNOVA

Stouthearted birders should fly up to Alaska for an unusual spring vacation. They may want to stay with Belle and Pete Mickelson at Goose Cove Lodge in order to experience what Belle calls "a world-class spectacular sight!" She says that during spring migration "we have 20 million shorebirds and waterfowl moving through over a month" in the Copper River delta. "My husband Peter figures that is the best place on the eastern Pacific Coast to watch spring bird migration." They suggest the last week of *APRIL* for the waterfowl migration. The shorebird spectacle is at its height between *MAY* 2 and 8, though Surfbirds, godwits, and some of the rarer species arrive a little later.

Spring and summer collide in Alaska and the birding sparks will fly. Catch this spring-summer marvel in the next chapter.

RUN FOR THE RARITIES

Coastal areas are particularly fine watch points for rarities. A Eurasian Green-winged Teal, larger and with slightly different markings than the Green-winged Teal of North America, was seen on Vancouver Island in *MARCH* and *APRIL* 1986. On a California pelagic trip in *APRIL* 1987, Solander's Petrels, far from home—ocean off Australia—caused some excitement. So did the Little Gulls, more at home on the other side of the Atlantic, than off Cape Hatteras Point in North Carolina.

Rarities make the avid birder's motor race and the east coast during the spring is a good place to roar when western palearctic strays drift in. The North American Rare Bird Alert keeps track of what's where. Bramblings, a Eurasian finch that winters in Juneau, Alaska, were seen in the Moser River National Seashore, Nova Scotia in early *MARCH* 1986 and in Ohio in *APRIL* 1987. Then there was the Common Chaffinch in New

Birders from all over the country show up when a rarity is reported

Brunswick in *FEBRUARY* 1986 and a Eurasian Goldfinch, that pretty little bird with a red face, was in the Hellertown, Pennsylvania, area early *APRIL* 1986.

APRIL a year later, birders celebrated a Bahama Mockingbird in Florida, Greater-Golden Plovers in Newfoundland, and a Northern Lapwing in Nova Scotia. Birding fanfares were sounded in Concord, Massachusetts, in early *APRIL* when a Fieldfare was sighted. Northern Wheatear, a handsome bird, was seen on Sable Island, Nova Scotia, in late *MAY* 1986. A year later in *MAY,* down at Buxton Point on the Cape Hatteras Seashore in North Carolina, some sharpies found a Spotted Redshank, a Curlew Sandpiper, and a Reeve all on the same day.

For *real* rare rarities, late spring-going-on-summer is an Alaskan "hot spot." Those Asian migrants bewitch otherwise sane birders into traveling to the very outposts of North America.

Four

Summer:
A Time for Nesting

Bluebirds called over us, and once a flock of goldfinches went roller-coasting by above the treetops. Time after time, clear above the rush of the stream, we heard the ovenbird set the forests ringing. But predominant among all this music of early summer there was that sweet, minor, infinitely moving lament, that voice of the north woods, the unhurried song of the little white-throated sparrow.

Edwin Way Teale
Journey Into Summer

*S*eals were resting on Hudson Bay's thick ice cover up in northern Canada and ice floes still swung back and forth on the tides at the mouth of the Churchill River. Mockingbirds were fighting dastardly intruders in hubcaps and windshields, dive-bombing maurauding cats. Ducks and sandpipers were scarce in southern waterways, and brown-legged bathers had replaced calling yellowlegs. Songbirds from south of the border had made it to Tennessee, Ohio, Montana, and Ontario. Trogons had already established families in old holes in southeast Arizona. And suddenly official summer has arrived and shorebirds have finished their nesting in the high Arctic and are beginning their long flight back south.

For most species, good old summertime is nesting time, one of the best times to see the "resting" nesters and to observe their often frantic feeding schedule. So is it a frantic time for some birders. And hot too. Sycamore Canyon in Arizona on the Mexican border has been known as "Birding's Bataan," but each spring and summer, hardy "hot shots" take the hot hike to look for the Five-striped Sparrow and, so close to the Mexican border, other exciting possibilities.

Not *all* summer birding is hot. You may freeze your tootsies looking for Ross' Gulls in Churchill, Manitoba, on Hudson Bay in *JUNE,* or trying to find the Siberian Tit in northern Alaska in *AUGUST.* Out on a small boat trip through the Head Harbor Passage from Maine over the invisible boundary to New Brunswick, *AUGUST* brought a cold rain that dampened all but the staunchest birding spirits. We donned all the clothes (six layers of T-shirts) we brought and were still cold. But between Deer and Campobello Islands we watched seals, Common Loons, Great and Double-crested Cormorants, nesting Bald Eagles, and rain.

Hot or cold, difficult or easy, summer birding can provide quality birding. It may be the only time of year you get off the home reservation and have the opportunity to visit another part of this magnificent continent. That is the time when Floridians take camping trips in the Canadian Rockies, when Tennesseeans visit Canada's Maritime Provinces, when New Yorkers head for Yellowstone, or Californians visit Maine. Summer is vacation time for many of us.

WHEN IS SUMMER?

Like other seasons, the boundaries of summer vary depending upon where you are. In subtropical south Florida, summer seems to last forever. It begins sometime in *MAY*, and extends to sometime in *OCTO-BER* or even into *NOVEMBER*. Some years, Florida is perpetual summer. At the top of the continent, the Arctic summer season is barely six weeks long. Much living must be packed into those six weeks! Mates must be found if such was not done while winging north. Nesting sites need to be refound and refurbished. Then laying, brooding, training, and traveling once again. In southeast Arizona, guide author Jim Lane identified a long summer: from the first of *MAY* until mid-*SEPTEMBER*.

Birders' summer is roughly from *JUNE* through *AUGUST*. Spring and summer are hard to tell apart in the highest latitudes and most of the birding "action" takes place at the end of *MAY*/first of *JUNE*. In a particularly cold year, someone may joke about summer having been "yesterday." Specific times in specific places are not as critical for the birder as they are during migration periods. Nesting birds aren't traveling birds. It is the birder who must travel.

As in other seasons, birds are found everywhere during summer. Perhaps only Mockingbirds and Mourning Doves are left over in the deep south, but in southern coastal areas from Brownsville, Texas, around the Florida peninsula, and on up north, terns and Black Skimmers will be nesting and certain gulls will be laughing their heads off along the beaches. For more serious birding, go north young birder, along with the other young birders of other ages. Alaska and Churchill on Hudson Bay in Canada are two favored summer "hot spots," cold though the temperature may be.

JUNE, the beginning of calendar summer, is the highlight month at the top of the globe, and is the pinnacle of the season all across the northern part of the continent. Birding activity is quieter in *JULY*, but already some species will be molting, donning more inconspicuous dress, and heading south.

NORTHERN LIGHTS

The *Aurora Borealis* is not the only sky spectacle during summer in the northern hemisphere. Skies in the northernmost reaches of the North

American continent are filled with the sights and songs of millions of birds. This is where the action is.

ATTACKING ATTU

Spring and summer are virtually one season at this outpost of North America at the tip end of the Aleutian Islands chain. It happens in late *MAY* through early *JUNE* on Attu Island, so far off into the Bering Sea towards Siberia that it's almost off the North American map. Dedicated, down-jacketed super-birders have been attacking Attu Island since 1976, some even sentencing themselves to annual visits. Birders have been going there for a late spring fling, or an early summer swing depending upon their point of view. Regardless, they are all there together. Larry Balch, president of the American Birding Association, pioneered birding trips there but says the sparse accommodations may not be available too far into the future.

A more inhospitable place is hard to imagine. So why would anyone of right mind, even an avid birder, go there? A look at the list of sightings spells the answer. It's the place to go for rare birds swept off course from Asia, the place to see some of the finest of North America's birding exotica. Do bear in mind that some world birders would rather see a Siberian Rubythroat in Siberia, and an Oriental Greenfinch some place in the Orient (and at that, the trip would not be much more costly). For those intent on seeing America first, and on building up their North American lists to dizzying heights of 700 or more, choose Attu. Benton Basham calls that far-flung island "the single most exciting birding experience" he's ever had.

Attacking the island each spring are bicycling birders equipped with walkie-talkies alerting one group to tear themselves away from admiring one rarity and hurry across rutted roads to another spot where another *very* rare species, such as the Spoonbill Sandpiper, has just been seen. Well, that one would really be worth pedaling hard for. In 1986, it was seen there for the first time in North America. It's rare anywhere but is more likely on the Russian side of the Bering Strait.

Some of the "foreign" birds seen at Attu are regularly but not commonly found in North America: Eurasian Wigeon, Tufted Duck, Green and Wood Sandpipers, Bar-tailed Godwit, and Ruff. Every year, others like the sandpiper with the spoon-like bill, make the list for the first time—front page news in the birding world. Bill Bouton and others on the Attu venture back in 1980 were tickled pink to see the Eurasian Dotterel, Terek Sandpiper, and Common Cuckoo in addition to other exciting rarities. In 1986, it was a Little Ringed Plover and a Red-breasted Flycatcher that captured top honors. A year later it was a Greylag Goose.

But 1988 flushed up some other "good" rarities reported to the North American Rare Bird Alert: Common Pochard, White-tailed Eagle, Mongolian Plover, Red-flanked Bluetail, Pechora Pipit, and the real highlight, a Yellow-breasted Bunting for a first North American record. With such "foreigners" visiting North American shores, why trek to Asia? Birds such as these would be relatively easy to spot in many places in Eurasia.

Some people just seem to relish hardship. Listen to Bob Odear telling about the 1987 Attu trip:

> To get to South Beach, you bicycle past lower base and through a place called Murder Point to the foot of an enormous, steep hill. A jeep road goes over the hill and you crest two peaks at about 600 and 500 ft. within a distance of approximately one mile. . . . We were still standing on the beach when a call came announcing that the continent's first Oriental Cuckoo since 1946 had been found. . . . Back over the mountains, back to home base, around Casco Cove, around the air base, past Debris Beach, (etc. etc.) . . . we pumped and stumbled. Six miles of agony! When we arrived at Henderson Marsh, we saw about 40 people already sitting on the hillside way off to the left. We collapsed on the tundra and watched the cuckoo. . . . I headed for home, elated with my new birds. Trembling from exhaustion and the pains in my legs, I forced myself to ignore the cold wind in my face and my still wet clothes. My last quest of the day was for a bed.

Yes, Attu is a tough place to go birding. Birders on their first trip there will add 40 to 50 birds to their life lists. Being young, strong, and agile as a mountain goat certainly helps, but most birders there are younger in spirit than their years would imply. Mary Davidson, former president of the Duval (Jacksonville, FL) Audubon Society (DAS) was 64 at the time she tackled the rugged terrain and her sister Peggy Powell, also a former DAS president, wasn't far behind.

ALASKA ATTRACTIONS

Many birders will eagerly visit mainland Alaska, a favorite summer vacation place. Alaska is huge. Distances are so enormous that travel by air is standard. It would take a generation of summers to see it all, what with 16 national wildlife refuges, 4 national parks, numerous other national forests and preserves, and more than 100 state park units.

JUNE, as early in the month as possible or even late MAY if you can do it, is the best time to visit Alaska if your objective is birding and not just general sightseeing and photography. Alaska offers superb birding throughout the summer, but the earlier you get there, the better your chance is of seeing some of the same straying migrants from Asia that the Attu attackers saw. If you really search, you might see a Bristle-thighed Curlew, Temminck's Stint, Common Black-headed Gull, or Northern Wheatear, and in places more easily accessed than Attu. How many birder's would like to add Whiskered Auklet, Bluethroat, or Short-tailed Albatross to their North American list? Alaska's the place, though nobody promises that such rare birds are readily seen.

Both land birding and sea birding are rewarding. Images of the remoteness and the spectacular scenery of the 49th state tugs at the toes of many a traveler. Birders know they will be seeing bird species shown in their field guides, many of them looking completely improbable. Perhaps the bait is puffins, those clownish seabirds whose enormous red and yellow parrotlike bill surely were created for some watery Halloween bash. Call it puffin seduction, it was puffins that drew me to Alaska that summer of 1979. It was the first time I had traveled to see a bird.

Although I had a fair-sized list of North American land birds (not a particularly neat list, I admit), I had never seen pelagic birds, those truly seagoing birds. In fact, those birds, puffins and auklets, on a middle page in my old Golden Press bird guide, looked so odd, it was hard to believe they were real. Who knows how many birders visit

Horned Puffin lures birders to Alaska

Alaska just to see those two species of puffins, the Horned and the Tufted Puffins? They breed in Arctic areas on both sides of the Pacific Ocean, arriving at their traditional haunts in late *MAY* or early *JUNE,* generally most of them all at once.

Do fly out to the Pribilof Islands if you can. That remote Aleut community in the gray Bering Sea is haven for nesting seabirds, enormous beach colonies of breeding northern fur seals, and heaven for those naturalists studying the wet arctic tundra. Our assemblage consisted of two distinct groups. As we spilled out of the old school bus at each stop, a group of young Japanese school teachers spread across the landscape down on their knees photographing dewy plants beneath huge camera lenses. Those of us in the other group, necks strung with binoculars, crowded together at cliff edge to spot puffins near their rocky high-rises.

Prominent island cliffs are summer breeding homes to one of the largest seabird colonies in the Northern Hemisphere. Thousands of Black-legged and a handful of rare Red-legged Kittiwakes, (nearly the world's entire population) alternately sail and sit. Thousands of Common and a few Thick-billed Murres turn black backs to the gray sea as they protect their eggs from rolling off the narrow ledge. Northern Fulmars fly incessantly across the cliff face. Red-faced Cormorants and several species of auklets, add to the wild cliff scene. Look down to the bottom of the cliff and you might see an arctic fox curled up asleep near water's edge.

On every beach, another spectacle greets the visitor to this distant island. A million, give or take a few hundred thousand, northern fur seals, their dark brown bulks almost indistinguishable from the dark brown rocky shores, snort like a million men gargling all at once. Aggressive "beachmasters," headmen of the multiple harems, keep constant tabs on their "ladies." The distinctive scent of so many seals crowded "wall to wall" wafts on the slightest breeze. At the diminutive end of the beach scene, puzzle over the darkish shorebirds and come up with Rock Sandpipers.

To catch even more of the flavor of international birding, go out to St. Lawrence Island or to Wales at the eastern tip of the Seward Peninsula, the point of the eons-past land bridge between Russia and North America. Not only do the Asian vagrants show up in large numbers but at Gambel, out on the western tip of St. Lawrence Island, thousands of migrating birds will be passing over at any given moment, headed for their high Arctic nests. Add all the eiders to your list in one day, and watch the rare and beautiful Ivory Gull overhead. Birds are not the only attractions, there are whales, whales, and whales.

On the Nome to Kotzebue sidetrip, favored by tour groups, the birding battalion will be looking for Old Worlders including, Slaty-backed Gull, wagtails, and Arctic Warbler as well as some of the North American high Arctic breeders. You will see many more birds if you go with a birding tour group rather than just trying to break away from the typical tourist group. Possibilities include Gyrfalcon, Brant, Rock and Willow Ptarmigan, Northern Wheatear, and White and Yellow Wagtails. Take a look at your field guide and see where so many species, particularly shorebirds, spend their short summers. Circling from Nome up to Pt. Barrow, over to Canada's Baffin Island and the Queen Elizabeth Islands, and on to coastal Greenland takes you halfway around the globe at the top end. Look at the maps in Harrison's *Seabirds* and grasp the enormity of the entire Arctic breeding area.

For birders with a passion for remote places, the high Arctic is sure to produce a real high. The Arctic, in contrast to Antarctica is not a solid land mass, but is a combination of ocean and islands. Two special islands in Canada's Northwest Territories might lure you if you fancy yourself a birding Peary. The huge Banks Island, way up above the Arctic Circle, boasts that its Bird Sanctuary Number One has the world's largest breeding colony of Snow Geese, along with large numbers of breeding Lesser Golden-Plovers and Buff-breasted Sandpipers. It is a veritable baby factory for Baird's and Pectoral Sandpipers, in addition to Sabine's and Glaucous Gulls. To the east, Bylot Island north of Baffin Island is another place to explore in early *JUNE* or in late *JULY*/early *AUGUST.* Those are the best times to see some of the same species. Watch also for Common Ringed Plover, Thayer's and Ivory Gulls, Snowy Owl, and King Eider. It's the lure of faraway places that brings a few intrepids to see these, and some otherwise familiar, birds on their breeding grounds.

Some birding groups schedule trips up there and some adventure tour groups such as Special Odysseys specialize in bringing visitors to this part of the world. You can expect to see loons including the Yellow-billed, all four eiders, jaegers, Snowy Owl, and a host of shorebird species including possibly a Rufous-necked Stint, Mongolian Plover and Curlew Sandpiper, and an occasional Ivory Gull. To this birdy list, Special Odysseys groups visiting the northern Baffin Island area can expect to see a handful of mammals: narwhals, beluga whales, polar bears, walruses, and seals.

Not every birder who visits Alaska is able to get to such remote places. Not to worry, there are areas more easily reached where the lust for Alaska bird species (and crabs) can be satisfied. Hop down to the Kenai Peninsula for some hot birding on a cool, sunny day. Tary in Potter's Marsh on the way down from Anchorage. From Homer take the ferry around the Kenai peninsula. Watch from ferry deck for anything that flies: both Arctic and Aleutian Terns, Common Eider, and both Kittlitz's and Marbled Murrelets. That is a first rate handful of lifers for many.

Rather than hanging over the rail on the big cruise ship plying the Inside Passage, birders want to take the boat trip into Glacier Bay to watch the calving of the glaciers. Only a "she-glacier" can "calve" and, accordingly, glaciers are referred to as "she"—"Thar she goes!" Puffins and other seabirds aren't the only animals you will see on your boat trip. Up on nearby cliffs will be mountain goat, and you will eagerly watch for the spout of a whale. Stay overnight at the lodge at Glacier Bay, dine on Dungeness crab, and walk through the silent, mossy rain forest and, at midnight, on the pebbly beach. That glow in the sky isn't the lights of a nearby city—there isn't one—it's the *Aurora Borealis,* the northern lights.

Alaska is famous for spectacular scenery. After exploring rugged coastal areas by boat, fly over gray, braided rivers, countless glaciers, and snow-covered jagged, rocky peaks including Denali, the highest mountain in North America. Hikers and birders will traverse the tundra and hike through alpine forests. Nonbirders will marvel at the herds of caribou parading across the ridge. Down the hill, a grizzly runs with unbelievable grace and speed. A fox carrying a rabbit in its mouth seeks shelter from an aerial attack—a Long-tailed Jaeger, perhaps a life bird.

The short spring/summer will enrich your bird list as well as your experience in one of North America's last frontiers. Watch for Spruce Grouse, the three ptarmigan, Glaucous-winged and Mew Gulls, Arctic Tern, Varied Thrush, Bohemian Waxwing, White-crowned and Golden-crowned Sparrows, and Lapland Longspur. Some of these birds still will be around in *AUGUST,* but that is the beginning of fall and many species will be leaving.

DOWN IN CHURCHILL

JUNE is also the best time to visit Churchill, down from the Arctic but not far down. Churchill is a tiny town at the mouth of the thousand-mile-long Churchill River. In summer, when the ice breaks up, the river empties into the west side of Hudson Bay, that enormous body of water in northern Canada at the latitude of southern Greenland. The town is dominated architecturally by the huge granary to which northern farmers used to send their wheat each summer when the ice pack on Hudson Bay would break up sufficiently for the wheat to be shipped out, principally to Russia. Each *JUNE,* Churchill is under siege by squadrons armed with binoculars, scopes, and cameras.

Because of its fame among birders, it is not uncommon to run into friends. John and Barbara Ribble, encountered the last time in Peru, made the trip here just to see the Ross' Gulls. Sitting at the next table in the dining hall were Andy and Joan Warren, friends from a couple of Africa trips. Admiring the Ross' Gull in the granary pond one evening after dinner was a whole group from the Hartford (Connecticut) Audubon Society, shipmates from a birding trip to the Dry Tortugas. You just never know who will turn up. Had you been with Victor Emanuel Nature Tours in June 1985, your fellow birder would have been Roger Tory Peterson. In recent years, Churchill has been the best place to see the Ross' Gull, a relatively rare gull and a relative newcomer to North American shores. It is not often found that far south of the Arctic Circle. Not only is it a rarity on this continent, it is also a lovely bird to observe. A small graceful gull with a black ring around its neck, its breast is often suffused with a delicate pink glow.

Mid-*JUNE* is preferred by the experts for this tundra territory. Anytime during that month is a good time to search out shorebirds, most in their breeding finery, including phalaropes, Stilt Sandpiper, and Hudsonian Godwits. Greater Scaup will be on the ponds, Lesser Golden Plover in the fields, while Trumpeter Swans fly overhead. Buntings and longspurs will be gobbling up seed on roads near the granary, and seabirds such as Parasitic Jaeger commonly fly across the mouth of the

Ross's Gull are Churchill's summer enticement

river. Common Eider, Red-necked and Pacific Loons rest on the river. Your bird guide may show "Arctic" Loon up there, but *Gavia pacifica* has been split from *Gavia arctica.* Willow Ptarmigan in mottled plumage scuttle away, flushed by an impatient rest stop seeker. Peregrine Falcons will be hard to miss as you unwittingly invade their nesting territory. Harris' Sparrow, giving forth its monotonous song, is not hard to see at roadside (Edward Harris was another of Audubon's buddies). Churchill is definitely a desirable place to fill in the blanks on your North American list.

Churchill claims to be the "Polar Bear Capital of the World" but you are not likely to see them at this time of year. Go back in the fall. Mammals won't be missing altogether for Ringed seals will be lolling far out on the Hudson Bay ice pack. Around the middle of the month, Beluga whales begin to come into the Churchill River, their ghostly backs looking like sleek icebergs.

Because Churchill is so remote, birding profits can be maximized by driving through Manitoba's boreal forest south of Thompson, an overnight train ride from Churchill. Occasionally, a birding tour group will spot a Great Gray or a Northern Hawk-Owl just ending nesting responsibilities. Forest, prairie pothole, and marsh areas between Thompson and the Winnipeg area offer heavenly habitat for *JUNE* birding.

Summer plumage of Willow Ptarmigan blends with habitat

Riding Mountain National Park with excellent, easy to reach birding opportunities is worth several days. Both the bird show and the animal show will produce memorable diary notes. At dusk, watch beavers building their everlasting dams.

Manitoba bestows a wide variety of birds: Sprague's Pipit, grouse, Gray Partridge (often found near the Winnipeg airport), flycatchers, and warblers. Southwestern Manitoba is part of that prairie pothole country of the upper midwest known as the "duck factory." Reaching from Alberta across to the eastern Dakotas and southeast into Iowa, this immense area is brooding area for ducks. Horned and Red-necked Grebes along with a wonderful collection of fuzzy ducklings trailing parent ducks on every pothole pond delight the traveling birder. Sadly, this area is rapidly shrinking in the race to plow the ponds. Black-backed and Northern Three-toed Woodpeckers, Boreal Chickadees, and Pine Grosbeaks are found in the boreal forests. Bluebird boxes nailed to fenceposts yield an increasing crop of Eastern and Mountain Bluebirds.

Excellent wetland habitat is easily reached north of Winnipeg at the Oak Hammock Marsh Wildlife Management Area. In addition to a variety of ducks, watch for Northern Harrier and Swainson's Hawk, Forster's and Black Tern, Marbled Godwit, and, with the help of a nice Swedish birder, you both conclude that the sandpiper is a Baird's Sandpiper. Some of the sparrows popping up and down in prairie grass will be frustrating but patience may help you sort them out.

Similar habitat and birds can be found in Northern Minnesota. Duluth, at the western tip of Lake Superior, was the site of the American Birding Association biennial convention in *June* 1988, so you know that the birding must have been worthwhile. It was. In addition to attending sessions on how to tell one female duck from another and other difficult bird identifications, members from the United States, Canada, and even Finland, chased about mopping up their list of needed lifers. High on the list of many was the Great Gray Owl. For at least one birder it was her 700th North American species. You may not give a hoot at such an accomplishment, but I literally hooted a Great Gray out of the dark forest where it was feeding an owlet. Many ABAers were after Connecticut and Mourning Warblers and found them easy to find. They perch and sing in summer, hiding in the underbrush the rest of the year. Singing Winter Wrens contributed all of their vocal artistry to the early summer morning concert.

TRANS-CANADA BIRDING

June is a favored month all across Canada. Canada's eastern provinces not only are favorite destinations for vacation-hungry tourists, they provide very special birding. Top of anyone's list is Bonaventure Island, two boat miles off Percé at the east end of the Gaspé Peninsula in Quebec. The gannetry on the steep rocky cliffs of the island is reputed to be the largest in North America. Between mid-*JUNE* and late *AUGUST,* your visual senses will be assaulted by more Northern Gannets than you thought existed—45,000.

The bird show in Gaspé, Newfoundland, and Nova Scotia will continue through *JULY* and even into very early *AUGUST.* If you have a time choice, early to mid-*JULY* would be best. Birding attention focuses on one of the highlights of a North American birder's life: a visit to the cliff-breeding birds of the northern coastland. From the Gaspé Peninsula east to Newfoundland then south to New Brunswick, Nova Scotia, and northern Maine, coastal and island cliffs are a madhouse of breeding activity. These maritime provinces are where landlocked birders may get a first taste of the excitement of pelagic birding without even getting into a boat. "In July we went to Newfoundland," Rob and Shirley Fine wrote. "That's where the puffins are. Would you believe 250,000 breeding *pairs.* We also saw gannets, murres, eagles, kittiwakes, assorted seagulls, plus whales. That cleared the wildlife agenda for us for a while."

Ferries in this part of the world are transportation links between the mainland and the islands, and between the islands themselves. In addition to getting you from one place to another, they provide opportunities to see pelagic birds of the North Atlantic. Some of the cliff birds can best be approached by boat as many will be found on islands. Most of the North Atlantic seabirds may be found nesting from Nova Scotia north to Greenland on this side of the ocean, and on the other side in coastal areas of northern England and Scotland, plus Iceland and Norway. Northern Gannets, Common Murres (cats purre), Black Guillemots, Black-legged Kittiwakes, Herring Gulls, Atlantic Puffins, and occasionally Razorbills will be seen crowded together on narrow cliffs sounding a cacophonic symphony.

JUNE to mid-*JULY* is the best time for adventurous birders to be in Newfoundland to see both nesting seabirds and forest birds. Make a foray to the French offshore islands of St. Pierre & Miquelon to see colonies of nesting Great Cormorants, Razorbills and Red-necked Grebes. Although the bird cliffs on Newfoundland are the premier performance, *JUNE* and *JULY* in coastal boreal forests from Newfoundland down to northern Maine provide an unbeatable opportunity for both sea and land birding. Many eastern warblers, perhaps as many as twenty species, along with flycatchers, finches, and sparrows nest in these pristine forests.

Machias Seal Island, the sovereignty of which is contested by Canada and the United States, is only accessible by private boat but the

Razorbills balance their nests on narrow cliffs

summer Atlantic Puffin nesting colony there is worth whatever diffi-
culties the trip may entail. Weather is often windy and rainy with seas
too rough to chance landing on slippery rocks. Early in *August* 1987, a
Wings, Inc. birding group was lucky: sunny, warm weather and calm
seas. The puffins were soon to leave, to disperse in the ocean as pelagic
birds do when they are not nesting. This day they were pecking and
preening and playing musical chairs on choice rocks. Viewing blinds
gave two marvelous pictures: the bustling seabird colony on one side
and, on the other, a nice wave of warblers stopping by for a bite to eat.

In this part of the world, ferries are not only an important transportation link, but incidentally provide the opportunity for easy
pelagic birding. Get good maps of the Maritimes and Maine to choose
your ferry route. If you haven't yet seen a Leach's Storm-Petrel, you
might be successful from the Bluenose Ferry running between Bar
Harbor, Maine, and Yarmouth, Nova Scotia.

On a Wings, Inc. birding excursion in early *August* 1987, our
group counted an astounding one thousand from the ferry deck. With
twenty-five hundred Wilson's Storm-Petrels pattering on the water, we
had an excellent opportunity to sort out subtle differences of "jizz"
between the two white-rumped storm-petrels. Wilson's and Leach's
Storm-Petrels are two of the world's 20-odd species of storm-petrels,
sparrow-sized pelagic birds most of which seem to skip across the

Atlantic Puffins nest on remote islands

ocean surface. Wilson's are thought to be the most numerous of all bird species in the world. They have been up in the North Atlantic for the summer and now are heading back down to Antarctica where they nest. Nesting Leach's Storm-Petrels will be found in some island rookeries. Scanning the waves from most any ferry, you will spot shearwaters, jaegers, and gannets, plus whales, seals, and porpoises. Ferries will also take you out to Grand Manan, an archipelago off the southern tip of New Brunswick, a birding "hot spot" identified in the Finlay bird-finding guide as having "everything." In early summer you will have the breeding seabirds, and by *AUGUST,* migrant landbirds will be coming through.

Not all summer birding experiences north of the 49th parallel occur in such relatively remote areas. North of Quebec, the Laurentian Mountains at Mont-Tremblant Provincial Park shelter a number of interesting birds. Probe for Yellow-bellied Flycatcher, Black-backed Woodpecker, Gray Jay, Boreal Chickadee, Olive-sided Flycatcher, Golden-winged Warbler, and both crossbills. If swallows are your thing, pass through Pembroke, northwest of Ottawa. Swallows, mostly Tree, are so famous there that there is a local Festival of Swallows during the time they peak the first two weeks of *AUGUST.* Mississauga and Oakville just north of Toronto offers excellent summer birding. From *APRIL* through *JULY,* try the native woods for Black-billed and

Yellow-billed Cuckoos, and in *JUNE,* look for Golden-winged, Blue-winged, Cerulean, Chestnut-sided, Mourning and Hooded Warblers, along with a Rose-breasted Grosbeak or two. Be sure to take mosquito repellent.

Crossing Canada by automobile is a very long trip. Mary and Tom Wood recommend birding from the vista dome of the Trans-Canada train. You would see more by leisurely poking around the pothole ponds, but the Woods saw plenty of ducks and shorebirds as well as a hawk or two and three species of blackbird.

Mid-*JUNE* can still be chilly and snow is likely at higher elevations in the Canadian Rockies, certainly one of the most scenic areas the North American continent has to offer. Driving up the Icefield Parkway between Banff and Jasper at the western edge of Alberta anytime during the summer is like driving through a postcard. For diversion, take the ice buggy across the Columbia icefield. Stop often on your route and enjoy noisy Clark's Nutcrackers and Gray Jays greeting you from tree branches just above your head. Varied Thrushes, Townsend's Warblers, Red Crossbills, Mountain and Boreal Chickadees, Three-toed Woodpeckers, and Bohemian Waxwings will be there for the looking. Climb a path up the mountain from the road and you might flush a White-tailed Ptarmigan. Easier to see will be elk, bear, mountain goats, and bighorn sheep at saltlicks beside the road.

Upland Sandpiper stands watch

Linger in the prairie country between the mountains and Calgary for it is alive with American Avocets, Marbled Godwits, and the colorful Wilson's Phalaropes spinning in pothole ponds. In the shortgrass prairie, check for Lark Buntings and Upland Sandpipers standing sentinel-like on fence posts. If you have time, head down to the Cypress Hills area in the southeast corner of Alberta. Sprague's Pipits, Baird's Sparrows, Burrowing Owls, and even Sage Grouse are possible.

Mid-*JUNE* into *JULY* is the best time for seeing nesting mountain birds to the west in British Columbia. You may see Northern Hawk-Owls, Lewis' Woodpeckers, and Williamson's Sapsuckers. Watch the cliff faces for White-throated Swifts and follow local directions to find hummingbirds, particularly the Calliope common in this area. Both scenery and birding during summer in this part of the world is soul-satisfying. To be sure of seeing the birds you want to see, invest in bird-finding guides. Finlay's *A Bird-Finding Guide to Canada* is most helpful throughout this magnificent land, and there are other guides to specific areas.

UPPER LOWER 48

In addition to the summer bounty Canada offers visiting birders, super scenery in the northern states combines with excellent birding. Summer lasts longer along the upper tier of states from Washington and Oregon across the continent through the Dakotas and the Great Lakes region to New England. Some nesting birds arrive in some areas in *MAY*, engaging in leisurely nesting activity until *SEPTEMBER*. Summer begins early out in Oregon and lasts longer. The Nature Conservancy offers several "Oregon Birdwatch" trips from *MAY* through *SEPTEMBER*. Your expedition will visit the coast, the mountains, and the desert, and will focus on the Malheur National Wildlife Refuge. Rare Trumpeter Swans nest on the refuge as do many ducks and hawks.

JUNE is a fine time to bird the American Rockies. The earlier in summer you are there, the fewer nonbirding tourists. Although considerably farther south than their relations in Canada, the mountains are higher, providing the birder with similar habitat. High up on Colorado's alpine slopes, still snowy in *JUNE*, White-tailed Ptarmigan are possible. Along the shores of alpine lakes, look in the tall trees for the Pine Grosbeak silently surveying nature's radiance. You could see the Three-toed Woodpecker and Red-breasted Nuthatch. Hard to miss in these mountains are Clark's Nutcracker, Gray Jay, Cassin's Finch, chipmunks, and golden-mantled ground squirrels. Keep your eyes open for both eagles, the Golden in more mountainous areas and the Bald near rivers and lakes.

Rocky Mountain National Park in north central Colorado is noted not only for these alpine species but also for birds of the Transition Zone. Some special ones are Williamson's Sapsucker, Mountain Chickadee, Pygmy Nuthatch, and Evening Grosbeak. Down in valleys cut eons ago by cascading rivers may be Hammond's and Dusky Flycatchers. In rushing mountain streams, it is the dark gray American Dipper that fascinates as it bounces from rock to rock and then into the water for a tasty morsel on the stream bottom. An abandoned drain pipe above a stream near Morrison provided a nesting and viewing opportunity to sharp-eyed birders.

While out west, naturally think of the pair of famous national parks, Yellowstone and Grand Teton in the northwest corner of Wyoming. Birders will arrive as early in summer as possible if only to avoid the crowds. Bubbling geysers in Yellowstone and glacier-clad peaks of the Teton Mountains are but the preamble to the scene that brings outdoor lovers back to this country time after time. Yellowstone might be known as geology in action; the Tetons as geological history. We now know that the great fire in Yellowstone during the summer of 1988, seemingly to headline scanners a disastrous event, was but part of nature's way of regenerating the vast forest, and the magnificent park rebounded with renewed life. Keep a mammal list as well as a bird list for these sister parks. Begin with bear, moose, elk, bighorn sheep, pronghorns, and beaver. Then watch the salmon climbing the spawning ladder at Jenny Lake in the Tetons.

Trumpeter Swans, reportedly 300 of them, also nest at Red Rock Lakes National Wildlife Refuge in Montana, fifty miles west of Yellowstone National Park. *MAY* through *OCTOBER* is the time during which this spectacle plays. A second avian marvel in this land of geologic splendors is the Whooping Cranes at Gray's Lake National Wildlife Refuge, Idaho, southwest from the Teton Mountains. The cranes are the second population of these magnificent birds and are the result of eggs brooded by Sandhill Cranes. This is the crane population that winters over in Bosque del Apache National Wildlife Refuge in New Mexico. To look at a wildlife map of Idaho, it appears that nearly the entire state is devoted to some kind of wildlife preserve: lots of birding habitat. Although noted for large collections of migrating waterfowl in both spring and fall, Idaho is well worth a summer visit if you're "out west." Some refuges boast substantial numbers of nesting summer ducks.

JULY and *AUGUST* are excellent months to visit Wyoming, where there are many state and national wildlife areas supporting a wide variety of bird life. Although many of these areas are well-known migratory areas, summer comes late in much of Wyoming. Pay particular attention to areas along the famous Green, Snake, and Platte rivers. Watch along the Green River for raptors and waterfowl. Western songbirds will be nesting nearly everywhere. Much of the area in the state is under the supervision of the Bureau of Land Management, but remember, some of it is private. Be sure to have detailed maps of the area and be careful about trespassing.

Montana has not been widely known as a hot birding state but in early *JULY* 1987, Four Points Nature Tours took binocular toters there. They were looking for Big Sky, Big Displays of wildflowers blanketing big prairies, Big Horns, Big Lake, Big Timber, Big Bison, and a bunch of other "Bigs." Montana offers a fine variety of western species particularly attractive to east coasters. Black and Vaux's Swifts would be noteworthy sightings. Varied Thrush, Calliope Hummingbird (smallest North American bird), Townsend's Warbler, grouse, longspurs, Baird's Sparrows, Three-toed Woodpeckers, Northern Goshawk, and Pine Grosbeak are all not only possible but probable. Montana has many excellent wildlife refuges where the birds you are looking for just might be looking at you.

Summer is a good time all across the northern states for seeing nesting songbirds and some nesting waterfowl. For many species of waterfowl using the Central flyway, the Dakotas offer both summer nesting and momentary resting sites that may be needed for those moving northward into Manitoba. At the edges of summer, very early in *JUNE* and late in *AUGUST,* try the Lacreek National Wildlife Refuge in South Dakota to catch some of the migratory species as well as numbers of nesting Trumpeter Swans, American White Pelicans, Upland Sandpipers, Black Terns, Burrowing Owls, Marsh Wrens, Warbling Vireos, Yellow-breasted Chats, Eastern and Western Meadowlarks, Orchard Orioles, and Swamp Sparrows. Stop there before or after going to Mount Rushmore and the Black Hills.

North Dakota may be lesser known as a tourist state, but it has some superb wildlife refuges. Summer birding activity is of variable length. Many of the migrant sparrows choosing the rolling prairies and the prairie wetlands for nesting sites arrive in *MAY* and stay until *SEPTEMBER.* Begin your summer early with a visit in *MAY* and you will not only find the summering nesters but spring migrant species such as Harris' and White-crowned Sparrows coursing farther north. Waterfowl stay around longer: from *APRIL* through *SEPTEMBER.* There's even a Lane guide to the state to open up your birding horizons. Visit the "Souris Loop" National Wildlife Refuges for special sparrows including the Sharp-tailed in early summer. Read Terry Rich's story "Land of the LGB's" (Little Gray Birds) in *Birder's World,* November/December 1987 issue.

Eastward from the Great Lakes region through to New England provides standard birding throughout the summer. Migrating shorebird stragglers seem to be northward bound through Wisconsin into very early *JUNE,* while southbound travelers appear later in the month. Early summer is the time to visit the Seney National Wildlife Refuge on the Upper Peninsula in Michigan, particularly while looking for Yellow Rails. As you will discover, they are not easy to find or see, generally requiring a night march through a marsh. While we're railing, down along the eastern coastal marshes you could look for Black Rails where there are several known nesting sites. Check with coastal wildlife refuges.

Nesting songbirds will be found throughout the northeast and in the eastern mountain areas. Keep looking for suitable habitat and take the time to stop, look, and listen. If you are having a Great Lakes summer, be sure to stop at the Erie National Wildlife Refuge in Pennsylvania just south of Lake Erie, and look for the Henslow's Sparrow, a real find for anyone's list. Mountain regions offer the summer vacationer so many treasures and there are so many people in the eastern United States to partake. Birders should be mindful of the hours of peak crowds and be up early in the morning to seek avian prizes.

Out on Cape Cod, that hook of Massachusetts south of Boston, tourists tend to crowd out birds and birders. Tourists who bring binoculars, walk the beaches, and peer into coastal thickets will be rewarded by a variety of eastern birds. Meander up the coast of Maine and visit Acadia National Park for a combination of forest and sea birding, and on up to Moosehorn National Wildlife Refuge on the border of New Brunswick to watch the American Woodcock, charmingly referred to as a "timberdoodle" in the evening in the blueberry fields from mid-*JUNE* to mid-*JULY.* Midsummer in Maine is the time for nesting Osprey and Bald Eagles. Watch in tall evergreens for Olive-sided Flycatchers. The woods will be full of vireos and warblers. "Isn't that a Blackburnian I see up there?" By *AUGUST,* Bobolinks and other summer nesters will be losing their breeding finery and by late *AUGUST,* many early migrants will be gone.

HOTTER "HOT SPOTS"

In the southwest, Arizona and Texas are still birding "hot spots" in more ways than one. Temperatures in much of Arizona will be in the 100s (the dry 100s) and at the bottom of the Grand Canyon, the heat is fearsome. Arizona and New Mexico are places in which to combine an interest in some of the most inspiring natural wonders of the world, the rich Navajo and pueblo Indian cultures, and of course western birds. If you are a real early birder, a family camping trip to the Grand Canyon may well be your introduction.

Mosey down to the mountains of southeast Arizona for some pleasantly cool weather. For those who thought spring offered the best of all possible birding worlds, veteran birder Bill Bouton says "Absolutely not! Nearly every species is equally easy to see in late *JULY*-first week in *AUGUST* when the excitement is doubled by the likelihood of post breeding dispersal of strays from Mexico!" He adds that *MAY* or *JUNE* is hottest (although a couple of Floridians thought it was downright chilly), but that summer monsoons of *JULY/AUGUST* "add cooling rains, spectacular cloudscapes, and shoulder tall green grass among the saguaros."

This is the season of the hummingbirds. With good guidance you should see Rufous, Violet-crowned, Black-chinned, Magnificent, Blue-throated, Broad-tailed, and Broad-billed with some ease. Experienced birders will be looking for Calliope, White-eared, and Berylline Hummingbirds, in addition to the Plain-capped Starthroat from south of the border. Two places to watch these tiny winged creatures at feeders are at the Southwest Research Station out of Portal and the Nature Conservancy's Mile Hi Ranch in Ramsey Canyon in the Huachuca Mountains.

If this is your first visit to southeast Arizona, you will be eager to check off those Mexican regulars: Elegant Trogons, Rose-throated Becard, Red-faced Warbler, and the nondescript Northern Beardless Tyrannulet, a real "LBJ" (Little Brown Job) or as some would call it, an "LGJ" (Little Gray Job). The Thick-billed Kingbird would be a fabulous "tick" as would be the Sulphur-bellied, Dusky-capped, and Buff-breasted Flycatchers. You will probably really work for Rufous-winged and Botteri's Sparrow but can't miss Bridled Titmice busily tending to their business in the branches in Cave Creek Canyon and other forested areas.

Southeast Arizona is superb owling territory. Listen to your birding tapes to help you identify the calls of Western and Whiskered Screech-Owls, Spotted, Flammulated, and the tiny Elf Owl. Madera Canyon is a favored spot for expert owlers. Walk around a lot at night, or better yet camp in places like Cave Creek Canyon. While waiting for the owl show, watch a family of Gambel's Quail scratching about in a garden at dusk.

Over in Patagonia (see *Run for the Rarities* at the end of this chapter), the Rose-throated Becard may still be there (it's never a certainty) and catch Black Phoebe, Bewick's Wren, and Phainopepla. Nearby, along the Sonoita Creek, you might see Zone-tailed, Gray, and Common Black-Hawks in the sky all at once. The Sonoita Creek Sanctuary, a Nature Conservancy preserve, is known as one of the best birding sites in the state.

JUNE in the Texas Hill Country is about as late in summer as you will find the very special Golden-cheeked Warbler from Mexico. It spends its nesting summer there but it's difficult to find it after it stops singing. Work hard for this bird and for another very special one, the Black-capped Vireo. Seek advice and check the where-to-find guides. *JULY* and *AUGUST* will find birding vacationers in Big Bend National Park, nestled into the big curve of the Rio Grande River. Stalwart birders will be looking in the Chisos Mountains for the famed Colima Warbler, found there from late spring through summer. This is its only known nesting area north of the border. This may also be the best place to see the Lucifer Hummingbird although southeast Arizona is a possibility. Watch for the Gray Vireo in the arid areas. More common birds will still please early birders: Band-tailed Pigeon, Common Bushtit, Hepatic Tanager, and Black-headed Grosbeak.

South Florida may not be your number one objective for a summer vacation, but it does have some advantages. It was cooler there than in much of the Midwest during the drought and heat spell in 1988. The beaches are safe for your children to play on, the prices for waterfront accommodations are lower, and you can pick up some special birds. Sanibel Island, and its famed Ding Darling National Wildlife Refuge, meet the above criteria. Named for political cartoonist and conservationist J.N. "Ding" Darling, the refuge on an island off the southwest coast of Florida offers excellent summer birding. Darling was the person responsible for the establishment of the annual duck stamp program that raises millions of dollars to preserve habitats for waterfowl. Today, with most National Wildlife Refuges requiring an admission fee, holders of the annual duck stamp will find free entry here. You will recognize one duck out there on the ponds of the "Ding," the Mottled Duck which lives here. It's a pale cousin of the American Black Duck.

For many birders and nonbirders alike, it is the gorgeous Roseate Spoonbills that get top billing. Often close enough to the wildlife drive for superior photographs, it is their colorful evening flight over the ponds to their roosting sites that evokes rhapsodic exclamations. Reddish Egret, often a difficult-to-find wader, is virtually a certainty any time of day, any time of year.

Not far away, endangered Wood Storks may be nesting in impressive numbers at the Audubon Corkscrew Swamp Sanctuary near Immokalee. Water conditions must be just right in the surrounding area or the nesting season will be a bust. Watch overhead for an American Swallow-tailed Kite gracefully swirling around, and listen along the boardwalk for the calling of the Barred Owls.

Key West, the southernmost city on mainland United States, meets our Florida summer birding criteria with charm and history to boot. Around the docks, pick out the Roseate Tern, recently placed on the endangered species list. A highly pelagic species, it comes in close to shore at this time of year. On your way down the Overkeys Highway, stop at the National Audubon Society Office in Tavernier and pick up a booklet telling about the better birding spots. Had you been there in *JULY* 1987, you too might have seen the Bahama Mockingbird that stopped by, and a year later, you might have seen the Zenaida Dove at the north end of Key Largo, a rare visitor from the West Indies.

JULY and *AUGUST* are hot in the center part of the state, but they are excellent times to see migrating shorebirds by the tens of thousands in flooded agricultural fields around Lake Okeechobee. Most anywhere in south Florida, bird spotters will see the wire-sitting Gray Kingbird, the Black-whiskered Vireo hidden in the bushes, Spot-breasted Oriole, and numerous resident natives and exotics. The Antillean Nighthawk can be seen at the Marathon and Key West Airports, and with diligence and the guidance of a local birder, the Mangrove Cuckoo can be found.

OFFSHORE

Early summer along both coastal beaches and shores supply many special species of interest although it is the pelagic birding opportunities in late summer that draw birders to opposite edges of the continent. Common shorebirds such as Sanderling and Ruddy Turnstone are up in Arctic breeding grounds, but other birds of shorelines have come up from wintering areas to the south. Least Terns are delightful but increasingly rare nesters along both lower coastal beaches. Tiny and ever busy, their high pitched calls are cherished sounds of summer for birding beach walkers. With increasing coastal development, however, suitable safe sites are at a premium and they have been known to lay their eggs on the roofs of shopping centers. The more spectacular Black Skimmer with its bright red bill, the lower mandible of which cuts the surface waters to skim up tiny fish, winters in South America but finds its nesting sites on the east coast, along the west coast of Mexico, and occasionally, as far north as San Diego.

Summer is not known for particularly notable birding in California, but because of the habitat diversity and high bird count, even random roaming of most any place in California should produce 100-plus species in a week or so. Flying in to take a pelagic trip, you will enjoy a few extra days in the state. For an easterner making a first trip to the west coast, birding excitement could rival the revival of the gold strike. Life birds are the nuggets and the early birder will gather a treasure trove.

AUGUST heats up pelagic trips off the coast of California. The last month of what we ordinarily think of as summer, is peak time, but keep at it through OCTOBER. Special sightings may be Black-footed Albatross, Buller's Shearwater, or Red-billed Tropicbird (oh, happy day!). Pelagic birding can be dreadfully dull out there with water, water everywhere, but these kinds of sightings, along with the whales and porpoises, can make a day on the water a stellar one in anyone's bird book.

Mid-JUNE through AUGUST brings glassy eyes to pelagic birding addicts aiming for the coastal waters of the eastern seaboard. They will be looking for Black-capped Petrels, Band-rumped and White-faced Storm-Petrels, Cory's, Audubon's, and Greater Shearwaters, Masked Boobies, Bridled Terns, and Great Skuas. There's no guarantee you will see any of these; just don't take the kind of seasick medication that makes you drowsy or you could miss one. Check the pelagic birding tour operators in the References at the end of this book and write or call for dates and departure points.

SUMMER WINDUP

Although much of the summer birding activity is concentrated in the northern reaches of the continent, southern stretches have a share of spots hot in more than one way. Summer, for many birders with families, is prime time for pursuing birds beyond the garden and nearby wildlife refuges. To be selective within your time considerations, look at the distribution maps in your field guide and make a list of the birds you most want to see. Then check out the state and national parks and wildlife refuges in the area. Ask for bird checklists; most have them. Thumb through *American Birds* and see what species were seen last summer. Some reports in that publication emphasize the rarities that invade 1 of the 24 regions covered, others provide a more even discussion of common species.

One wouldn't necessarily think of the popular tourist objective, Mammoth Cave National Park in Kentucky as a particularly pleasant birding area, but if you are filling out your list of wood warblers, take a look at what you might find. Summer visitors include many eastern warblers: Prothonotary, Worm-eating, Blue-winged, Northern Parula, Yellow, Cerulean, Yellow-throated, Prairie, Hooded, and Kentucky, along with Ovenbirds, Louisiana Waterthrush, Yellow-breasted Chat, and American Redstart. Crowded campsites in North Carolina's Smoky Mountains will yield many of these same summer visitors. Rhododendron will be blooming along virtually every mountain roadway until early AUGUST, adding an extra dimension of viewing pleasure.

AUGUST signals summer's ending. Visitors from the south are packing it up. Fall migration is well along for many species in many places. This popular vacation time provides ample opportunities to watch the fledglings, observe the molting of old for new flight feathers, and observe the relentless movement south, but we've just skimmed the surface.

RUN FOR THE RARITIES

As in other seasons, bird species that seldom are, or even never before have been, seen in North America are reported and viewed by dozens to hundreds of birders eager to add one more to their North American list. To get an idea of what happens, listen to this from the North American Rare Bird Alert (NARBA) Monthly Newsletter of August 1987:

> Speaking of expressions that birders use, the 'Patagonia Rest Stop Effect' (a phenomenon in which lots of good birds are found in a certain spot because so many birders are there looking for the good birds that have already been found there – and on and on) is the one that best describes what happened in July – not in Arizona – but just outside of Dover, Delaware. A White-winged Tern was found at the NW corner of the Port Mahon Rd. impoundment in Little Creek on the weekend of July 11. NARBA subscribers from as far away as California, Ohio, and Georgia (as well as DE, MD, PA, NY and NJ) were on location as early as Monday and Tuesday. The bird was still there later in the week when the Washington hotline began carrying the report. Birders combing the impoundments and nearby Bombay Hook refuge found Ruffs and Reeves, brightly plumaged Rufous-necked Stints and Curlew Sandpipers!

Were you there? To elucidate for the early birder, the tern is an Old World tern that looks much like the North American Black Tern; the Reeve is the wife of the Ruff; the stint, an occasional visitor to Alaska from its Siberian breeding grounds, is a *Calidris* sandpiper, a nearly worldwide group of spotted sandpipers and close relative to some of North America's such as the Western Sandpiper; ditto for the Curlew Sandpiper. Knowing something about families of birds adds some zip to birding.

Alaska is always getting lots of rarities, but it's a little far for most chasers and twitchers. Closer to home for most of us were some of the following rare bird sightings reported by NARBA in the summer of 1988: Garganey in Vermont, Yellow Rail in 3-M country (Manitoba, Michigan, and Minnesota), and Black Rail in North River Marsh near Beaufort, South Carolina; Red-billed Tropicbird and Gay Head Cliffs at Martha's Vineyard, Massachusetts; Masked Booby at Gulf Shores, Alabama; Fork-tailed Flycatcher, way out of its Mexican bounds, near

Williamsburg, Virginia; Bar-tailed Godwit at Chatham, Massachusetts; Yellow-green Vireo at Laguna Atoscosa National Wildlife Refuge, Texas; Terek Sandpiper at the mouth of the Carmel River in California, to mention a few.

New rarities turn up every year; other rare and hard-to-find species show periodically. The year before, NARBA reported *JUNE* jewels you may have missed: Bridled Tern off Key Largo, Florida; Eurasian Jackdaw in Lewisburg, Pennsylvania (what *did* it do to get into the maximum security prison?); and Fan-tailed Warbler and Five-striped Sparrow in Sycamore Canyon, Arizona. The pesky Eurasian Wigeon (pesky to find, that is) showed up in two famous birding refuges, Plum Island near Newburyport, Massachusetts, and Jamaica Bay, spittin' distance from New York's Kennedy International Airport. *AUGUST* was a fine month for owls. Flammulated in Arizona both *JULY* and *AUGUST,* Northern Hawk-Owl up in the Yukon Territory and in British Columbia; Ferruginous Pygmy-Owl in Dudleyville, Arizona; Boreal in Routt National Forest, Colorado; Northern Saw-whet in Rustler Park meadows in Arizona's Chiricahua Mountains; and a Wag, a Black-backed Wagtail, was spotted in Ventura, California.

You mean you were birding in the area and didn't know about these fine sightings? Subscribe to NARBA so you won't miss the excitement again. It's absolutely amazing what the excitement of the chase does for the heartbeat, and for the spirit. Try it some time.

Five

Fall:
A Time to Move South

... in the waning summer nights, bobolinks were taking wing for far-off South America. These were the early ripples of migration, ripples we would later see mount into the great waves of the autumn flight.

Edwin Way Teale
Autumn Across America

*F*all is the flip side of the migration coin. Birds that in the spring flew north on wings of song to summer nesting sites, have donned new flight feathers and, with progeny, return now flying for the most part in a southerly direction. Fledgling geese and goshawks, Veerys and vireos, warblers and wigeon, terns, turnstones, and tanagers are venturing forth on their first treacherous trip to winter homes. Many will not make it. A few oceanic visitors to northern waters did not breed. Wilson's Storm-Petrels and South Polar Skuas fly south not to rest but to nest in the austral summer of Antarctica. Sooty Shearwaters will nest short of that icy continent at the bottom of the globe. Not all birds travel. Resident species will still be about. Most owls are stay-at-homes. Chickadees care little for wandering. So if you miss the migration traffic, get out and watch the rich reds and golds of autumn, enjoy the cool clarity of the air, and maybe visit the polar bears in Churchill.

WHEN IS FALL?

"Fall is in the air" for many birds, as early as *JULY*. From the high Arctic to the sweltering agricultural areas of southern Florida, there is evidence of the restless movement of migrating birds. In the high Arctic, summer is very short. Top end nesters must do their thing and leave quickly. Sometimes their food supply, those juicy insects that so bedevil the birder, have finished their cycles. Some birds just seem to know when it's time to go.

Like other seasons, fall comes at different times in different places. Birders' fall is considered to be from *JULY* through *NOVEMBER* although it will be difficult to persuade a traveler through Mississippi that it's "fall" in *AUGUST*. A useful but fallible generalization would have us believe that the farther north, the earlier the fall. If you think like a shorebird, point your little beak south in late *JUNE* or early *JULY*. What birders generally think of as fall, however, runs from about mid-*AUGUST* through *NOVEMBER*.

By mid-*AUGUST*, the traveler as far south as northern Vermont is aware that warblers and other migrating species are becoming more abundant as they stop along the way to feed. Leaves are beginning to trickle to the ground and a hint of coolness is in the air.

It is in the northern part of the continent that both the weather and the movement of birds signals the beginning of the fall migration. Just when fall happens where depends in large part on weather. From a birding viewpoint, the migration may sort of evaporate at a given time in a given spot. Reporters for *American Birds* may report for a particular area that fall birding was "lackluster" or that the hurricane season along eastern coastal areas was "disappointing." That means there were no bad storms to bring in the "interesting" species. Water levels in the many refuge lakes and ponds often determine whether or not shorebirds or waterfowl touch down at those places. Birders may sometimes conclude that fall just "flies over."

As far south as south Florida, migrating shorebirds are seen as early as *JULY*. In fact, the heat of the summer is the time to "think cool" when visiting the flooded fields around Lake Okeechobee. Great agricultural combines are preparing vast areas for fall planting and winter harvesting. Their flooded fields attract all manner of waders and shorebirds by the hundreds of thousands. Even an occasional flamingo finds its way to this shallow soup.

At the other end of the fall migration is *NOVEMBER* in Niagara Falls, which, for anyone other than a zealous birder, is not going to be any more a favorite time and place than is central Florida in *JULY*. For the gull-fancier, however, it's the perfect point to be around Thanksgiving. Millions of gulls, give or take a hundred thousand, follow the Niagara River through that narrow gorge connecting Lake Ontario with Lake Erie. It's a chance to grasp an identification handle for many species of gulls in their many plumages. Field Guides, Inc. trips have clocked up to 13 species in a single day. Just about every gull that flies through the eastern part of the continent will be part of that fall migration party.

No one knows how many birds migrate but estimates run into the hundreds of millions. *National Geographic* magazine in a major article on migration, furnished estimates for waterfowl in North America (August 1979). Of the estimated 100 million ducks, geese, and swans that leave their North American breedings grounds at the end of summer, the *National Geographic* article estimated that only 40 million would make it back the next spring. Waterfowl and other long-distance migrators fly, sometimes nonstop, across Caribbean, Atlantic, and Pacific waters to their selected winter homes. Some of those wintering locales are in more southerly parts of North America, while some are in Central and South America—as far south as Argentina. For some species, this means flying from as far north as the high Arctic—nearly incomprehensible distances.

Miracles are many. To think that the Arctic Tern, all of 15 inches long, makes the 25,000 mile annual round trip from Arctic to nearly Antarctic strains even a Star Wars imagination. The Blackpoll, a black and white wood warbler, is one of several birds which travel nonstop for 2300 miles in 86 hours. That's nearly 27 miles an hour. Greater Shearwaters, often seen on a North Atlantic pelagic trip, wing 8000 miles across the Atlantic Ocean to a tiny dot on the map, the Tristan da Cunha island group lying between South American and the southern tip of Africa.

Most places in North America where these birds stopped on their spring trek, are places to watch for their return visit in the fall although many warblers and other small birds have quite different fall routes. Southbound Chestnut-sided Warblers generally fly decidedly east of their northbound patterns. Accordingly, some places are better in the spring for migrant watching, others in the fall. Songbird concentrations seem more focused in the spring, more diffuse in the fall. Or perhaps we just don't have as many auditory clues about the bird's whereabouts in the fall. Shorebirds moving northward along the coasts in the spring, will generally be found moving southward along those same coasts in the fall.

Waterfowl and hawk passages in many areas are somewhat spread out in the spring, but more concentrated in the fall. There are greater numbers in the fall, too. Any birder who, during the fall migration, has watched parent ducks followed by their many ducklings can attest to this. Lots more fly south in the fall than return in the spring. The smaller returning flock is due to many reasons: hunting, predation, disease, and exigencies of old age.

Like those in the spring, fall birding "hot spots" have been ferreted out by eager birders. Follow mountain ridges, which provide such wonderful updrafts, for the best hawk and eagle flights that sometimes occur in astounding numbers in the fall. Follow down the Mississippi River which provides a guideway for millions upon millins of waterfowl. Follow that heavenly Central flyway down through which the Whooping Cranes fly into south Texas. Watch almost anywhere along broad coastal areas as well as offshore.

HAWKING "HOT SPOTS"

For keen birders, fall is dominated by the hawk migration. Blown by northwest winds, raptors soar by the tens of thousands towards eastern coastal waters. Official hawk watches have been established at places such as Hawk Mountain, Pennsylvania, and Cape May, New Jersey, two of the more famous ones. All around suburban New York, hawk watch totals are checked by birders just as the stock market totals are checked by bankers (by some birders, too!). Up and down the East Coast throughout the fall, from the Maritime Provinces to Key Biscayne, hawk watching is fun. Bruce Neville, vigorous birder and veteran Key Biscayne hawker, suggests that the Key is the best place in all of North America to watch Peregrine Falcons chasing parrots. Now how about that for fun?

Pick your spot and watch. A sampling from *American Birds* for the fall of 1986 goes something like this:

- Broad-winged Hawks, (30,535), *SEPTEMBER* 14, Quaker Ridge, Connecticut
- Broad-winged Hawks, (30,000), *SEPTEMBER* 14–15, Upper Montclair, New Jersey
- Peregrine Falcons, (117), *OCTOBER* 4, Assateague Island, Maryland/Virginia
- Northern Harriers, (15), *OCTOBER* 16, Cape May, New Jersey
- Northern Goshawks, (5), *OCTOBER* 11–12, Brier Island, Nova Scotia
- Golden Eagles, (6), after mid-*OCTOBER*, Quaker Ridge, Connecticut
- Short-tailed Hawks, (2), *OCTOBER* 27, Big Pine Key, Florida
- Bald Eagles, (23), fall, Quaker Ridge, Connecticut
- Sharp-shinned Hawks, (1000 + on some days), Lighthouse Point, Connecticut
- Osprey, (789), fall, Hawk Mountain, Pennsylvania
- Merlins, (78), fall, Hawk Mountain, Pennsylvania

A year later, *American Birds* reported 22 Swainson's Hawks from *OCTOBER* 6 to *NOVEMBER* 24 in the Florida Keys, and 8 of the elusive Short-tails on *NOVEMBER* 8 in the Everglades National Park.

CAPPING CAPE MAY

Mid-*SEPTEMBER* to mid-*OCTOBER* is peak action time at Cape May, New Jersey. From *SEPTEMBER* to as late as *FEBRUARY*, you will find an ample supply not only of raptors but of many other migrating species as well. As many as 200 species are reputed to have flown through in a single day at the height of the fall migration. Perhaps by now, that little finger of New Jersey that sticks down into the Delaware Bay will have been designated as a National Wildlife Refuge. It is an essential area for both breeding and migrating birds and is a staging area for huge numbers of hawks and other raptors. These magnificent flying machines suddenly find themselves at the end of the long, narrow coastal plain. Caught between the Atlantic Ocean and the Delaware Bay, and being reluctant to fly over the ocean, they change course and fly up around Delaware Bay before proceeding again down the coastline.

Dunes, with ocean on one side and coastal thickets on the other, ponds, marshes and meadows, weedy fields, woodlands, beach, and tidal flats, together offer the varied habitat needed to attract the variety of species which makes Cape May such a thrilling place for the binocular gang. Finding the action at a given hour on a given day won't be difficult. The grapevine is nearly instantaneous. Finding out about hawks is as easy as listening.

- Broads ketteling high to the left.
- Here come the Sharpies.
- Osprey overhead.
- *Great* day for Peregrines.

The conversation is all about hawks. Early birders struggling with often hard-to-identify hawks can count on friendly help from eager "experts." One hawk watch veteran volunteered the jingle about the Sharpie, or Sharp-shinned Hawk:

Flap, flap, sail
Long square tail

Amazing how many sharpies one sees after that.

Listening intently, you might think your companions, who come from all over, are talking about hospitals when they mention what sounds like "Humana." They are not giving hot tips for the stock market, but talking about the Hawk Migration Association of North America (HMANA). They all seem to belong to this fine organization and you find yourself asking how you too can belong.

Get to the official hawk watch platform at the Lighthouse just south of town by sunup. Paul Beach from down Wilmington way does. "It's a tradition. I like the feeling of being here in the darkness just before dawn listening to the soft sounds of the marsh. This is a yearly pilgrimage." Each day brings forth new species, more or fewer than the day before. You will keep track of the counts posted on a large bulletin board. OCTOBER 5, 1987, featured the Broad-winged Hawk

Hawk counts provide daily news at the Cape May Hawk Observatory during fall migration

show. Official counters from the Cape May Bird Observatory counted 1400 in wave after wave, kettle after kettle. The count board showed for that day:

164	Cormorants
785	Sharp-shinned Hawks
14	Red-tailed Hawks
14	Red-shouldered Hawks
1400	Broad-winged Hawks
1	Golden Eagle
125	Northern Harriers
149	Osprey
29	Peregrine Falcons
22	Merlin
304	Kestrel

for a total of 3019 individual birds flying over the Cape May Observatory. Three days later, during a lull between aerial waves, one young expert, walking across the parking lot, suddenly shouted, "Long-billed Curlew." Fifteen binoculars pointed skyward as the bird called out its name. That is a pretty exciting bird on an eastern hawk watch. It reminded me of the Common Loon I saw from a Hawk Mountain watch some years ago.

Total numbers racked up in a single day can be pretty incredible. Official count for OCTOBER 2, 1986, was 12,064 individuals. Sharp-shinned Hawks and American Kestrels led the list. As if that wasn't enough, how about the thrill of seeing 140 Peregrine Falcons that very day? Many of us count ourselves lucky to see a couple a year. A year later the HMANA count for Osprey was 5402.

When you tire of looking out across the marsh grasses or when word arrives of a warbler wave in the woods at Higbee's Beach, jump in the car with newfound friends and head over there. These "wave days," when the migrant songbirds are surging in substantial numbers, will really boost your list. Even if a nonlister, it's exciting to see so many species and so many individual birds beating the bushes to extract every available berry and bug as they prepare to continue south, sometimes crossing wide waters.

If it's a chilly day, visit a charming restaurant in this picturesque town for hot coffee and fresh baked breads. Cape May, at the southern terminus of the Garden State Parkway is a vigorously Victorian town, one that would be worth visiting just to sample the flavor of a bygone era. Cape May is full of friendly birders and friendly residents. An automobile mishap just north of town failed to alter my hawk watching schedule. Dolly Nash, retired physical education teacher in the local school system, rescued me and then taxied me around to all the best birding sites, introducing me to local guru and national hawk watching expert, Pete Dunn. Binoculars hanging around your neck is better than a fiver in your hand when it comes to hitching rides.

Some years there are larger concentrations of birds than others; some days are better than others. If at all possible, plan to spend a long weekend, or even a week, to experience the full impact of this spectacular place. Between Cape May and nearby Brigantine National Wildlife Refuge an assortment of fall migrants gives excitement to birding. Snow geese fly in lovely lines across the marshes from OCTO-BER through early NOVEMBER to land wave after wave in the ponds at Brigantine. This is a favored resting place on the last leg of their migration from Baffin Bay in northeast Canada to wintering areas as far south as the Texas coast. Mute Swans swim there too. They really belong on the other side of the Atlantic, but in a few locations in north-eastern states, they have become sufficiently established to be included on the official North American checklist. You can count them here, but don't count those gracing some city park pond.

Tree Swallows swirl in huge clouds of frantic activity at this time of year. In nearby woods at both locations, Eastern Phoebes, kinglets,

Yellow-rumped Warblers, litters of Gray Catbirds, wrens, and sparrows perform. Each day's show is different. White-rumped and Baird's Sandpipers will be spotted and sometimes there will be a bell-ringer. Many scopers had an unexpected bonus at Brigantine in OCTOBER 1987, a Curlew Sandpiper.

In 1986, hawk watchers saw a surprising Black Rail on SEPTEMBER 15. You might wonder how to watch hawks in the sky at the same time you are seeing a rail scuttling through marsh grasses. Well, Cape May is a hawk watching spot where you just look *across* the marsh towards the town on the near horizon. Legend has it that a Black Rail even showed up one time in the town's little shopping mall.

Offshore scoters, Forster's and Royal Terns, and an occasional flock of Northern Gannets flying low over the ocean are regulars. An east wind could bring in jaegers and who knows what rarity?

Fall take-off points like Cape May have the same considerations as do spring landing points: weather and water. A storm can bring birds down by the thousands. Points of land along major lakes or seas often see great gatherings of restless birds seeming to screw up their courage for the take-off. Actually, these birds are busily feeding, building up their inner resources.

American Birds, in the Spring 1987 issue, carried a report from regional editors Robert O. Paxton, William J. Boyle, Jr., and David A. Cutler on that part of the world the previous fall:

> "Migration started off with a bang. Precocious cold fronts brought down a lot of passerine migrants in August. . . . September and October brought calmer weather and, some thought, dull birding."

. . . The Wings, Inc. birding tour catalog says, "Given northwest winds, Cape May can provide the most exciting fall birdwatching in North America."

HAWKING THE MOUNTAIN AND OTHER HAWK SITES

Just the opposite of "dull" was the Paxton/Boyle/Cutler report the following year for Hawk Mountain. Had you been there on OCTOBER 1 you would have witnessed flights of 14 species of hawks and both vultures, "the greatest daily variety in this watch's half-century." Hawk Mountain north of Reading in eastern Pennsylvania is one of the most accessible for eastern birders and is one of the most popular. The Hawk Mountain Sanctuary is the granddaddy of all the hawk watches and is today one of five official hawkwatches in the northern Appalachian Region. Don't expect record-breaking waves every day or every year but your visit there should be one of your birding year's highlights.

Hawk watchers have valley vista at Hawk Mountain

Pack your lunch and a warm thermos for it tends to be chilly in *OCTOBER* on mountain lookout points. You will have plenty of company, particularly on the weekends and there will always be sharp-eyed watchers readily identifying the different species. With luck, a steady stream of dark spots looms on the horizon over the Kittattiny Mountain Ridge, part of these ancient Appalachian Mountains. Quickly the dark dots become recognizable as raptors, then as Osprey, Broad-winged, Red-shouldered, Rough-legged, and other hawks, along with Merlins and a few Peregrines. There's leeway on time to be there. Veteran spotters report that "Broads" begin moving through in late *AUGUST,* peaking in mid-*SEPTEMBER.* The best time for most species is late *SEPTEMBER* through *NOVEMBER.* Northern Goshawks and Golden Eagles generally peak from late-*OCTOBER* to mid-*NOVEMBER.*

Even in suburban New Jersey, you can put binoculars in your brief case and catch major movements through Montclair. Peak days are different for different species so keep your eyes on the sky in the fall. Peak 1986 daily counts at the Hawk Mountain Sanctuary, according to the Hawk Migration Association of North America, were 583 Sharpshins on *OCTOBER* 4 and 15,745 Broadwings on *SEPTEMBER* 14. But the prize goes to Greenwich, Connecticut, where watchers counted twice that many Broadwings that day. What spectacular sky kettles there must have been!

Point Pelee, best known for its spring influx of passerines, is also a well-known fall migration point, particularly for hawks. Fall migrants begin to drift through this point of Canadian soil jutting into Lake Erie south of Detroit as early as late *JULY* some of the shorebirds are heading south. They go through in sizable numbers from mid-*AUGUST* to late *SEPTEMBER* or early *OCTOBER*. Mid-*AUGUST* is known for flycatchers, and from then to mid-*SEPTEMBER*, waves of Rough-winged Swallows will fill the skies. *SEPTEMBER* to mid-*OCTOBER* is considered the peak period for ducks. Gray-cheeked Thrush should be easy to see in late *SEPTEMBER* and Water Pipits in *OCTOBER*.

American Kestrels abound from late *AUGUST* through mid-*OCTOBER* and Ospreys during most of *SEPTEMBER* and *OCTOBER*. Late *SEPTEMBER* is favored for Peregrines and Merlins. You should see Northern Goshawks, Sharp-shinned, Cooper's, Red-tailed, and Broad-winged Hawks. Rough-leggeds begin to come in large numbers the latter part of *OCTOBER* and stay through the winter.

A few of the warblers will be easily seen during fall migration although warblers warble best in Pelee in the spring. Canadas (warblers not geese) are generally around from mid-*AUGUST* to mid-*SEPTEMBER*. *SEPTEMBER* is prime time for Black-and-whites, Tennessees, Magnolias, Cape Mays, Bay-breasteds, and Wilson's. White-crowned and White-throated Sparrows are around most of the fall.

This variety of dates illustrates the difficulty of pinpointing the exact time to find a select species at a predetermined point. Best advice if you travel long distances to one of the "hot spots" is to spend several days or even a week or two in a general area known for its migration. While there, check out the resident species and visit the local points of interest.

The northeast United States is not the only place where hawk watchers watch hawks. It is, however, the area that has the most action. If you lived in the St. Louis area and loved Broad-winged Hawks, you would be happy as a harrier. Late *SEPTEMBER* to early *OCTOBER* seems to be peak season and a HMANA official watch there reported 11,543 overhead on *OCTOBER* 4, 1986. The next year you wouldn't have been so happy. Hawks funnel down the North American continent to tiny Panama, the land bridge to South America where they are seen in unbelievable numbers. There's a HMANA official watch there and by *OCTOBER* 26, watchers clocked 184,400 Broad-wings, 179,620 Swainson's, and 179,620 Turkey Vultures. During the southward passage that year, watchers recorded just under a million Broad-wings and just over a million Swainson's Hawks winging their way to Argentina.

Join a local group doing an official hawk watch and get some expert guidance on identification, or write to the Hawk Migration

Association of North America (see Appendix 2 at the end of this book for the address).

AMBLING ALONG THE ATLANTIC

East coast fall birding can be of interest even if you are not a raring raptor-watcher. For a major fall migration trip, you might begin in the southern portion of Canada's Maritime Provinces as early as *JULY* and work your way south. Cliff-nesting birds will be finishing their domestic duties and in *SEPTEMBER* you will see a few farther south on a pelagic trip. Shorebird passage will run from then to mid-*SEPTEMBER*; and until late *AUGUST* for landbirds.

John Leo, who wrote the *Time* magazine story on the popularity of birding, tried a late *SEPTEMBER* pelagic trip farther down the coast. Writing in the January/February 1988 issue of *Birder's World,* he reported some of the same petrels and some shearwaters too—Manx, Greater, and Cory's. An earlier *American Birds* sponsored pelagic trip spotted a White-faced Storm-Petrel, a starred species on most "want lists." Fall is a time to add to your "landbirds seen on pelagic trips" list. Leo listed a couple of dozen of the southbound migrators winging over the Atlantic. They sometimes fall exhausted on anything that floats, including your boat!

Landbirds will be passing over land as well until mid to late *OCTOBER*. Early *SEPTEMBER* flycatchers, warblers, and vireos are moving south through spruce forests. Throughout these coastal areas until mid-*SEPTEMBER,* shorebirds on migration can be seen on the many flats at low tide. Take your down jacket and pray for a cold front to give you the best show. Listen for the Boreal Chickadee along the path through the spruce woods, and peer expectantly through the dark undergrowth for the elusive Spruce Grouse.

ISLAND HOPPING

Island hopping down the East Coast from *AUGUST* through *NOVEMBER,* sounds like lots of fun and some years provides impressive birding, though it may vary from place to place and year to year. Here's a sample time frame.

Mid-SEPTEMBER

Monhegan Island off the coast of Maine, a well-known migrant "trap," is worth the price of the ferry trip. Once there, you won't have difficulty finding the birds. It's a small island eagerly sought by both birds and birders. Even for those who know well the typical eastern migrants partying there, the island's isolated location well off the middle coast of Maine is a choice spot. A landing target for all manner of strays, you could see a Magnificent Frigatebird, well north of its usual southerly soaring circle. In other years, birders have spotted Le Conte's and Henslow's Sparrows (sharpen your eyes), and such western species as Lark Bunting, Say's Phoebe, and Rufous Hummingbird. You might even see that arctic breeder the Northern Wheatear, much more common on the other side of the Atlantic Ocean. Rare birder Bob Odear warns, "There's no real reason to be on Monhegan Island for birding unless a west or northwest wind has blown." When that happens, he says it could "rain passerines."

Mid-SEPTEMBER to Mid-OCTOBER

Martha's Vineyard, is another island "hot spot," highly recommended by the bird-knowledgeable. *American Birds* regional editor Richard A. Forster reported a Red-billed Tropicbird was a startling visitor to that well-known vacation island off Cape Cod on SEPTEMBER 15, 1986. You just never know what good or ill wind will blow in an exciting bird. Massachusetts, a state known in the bird world for both birds and birders, is fine for fall migrants. Watch for heavy concentrations of waterfowl at the Great Meadows National Wildlife Refuge not far from Boston. And don't forget that many of the migrants that visited Plum Island up near the New Hampshire border in the spring, will be back in the fall.

Early OCTOBER

Follow up Monhegan with a visit to Block Island off the coast of Rhode Island. It's another one of those fall fiestas that may result in upwards of 150 species in one weekend. Although late SEPTEMBER is alright, members of the Rhode Island Audubon Society recommend the first weekend in OCTOBER and you can go out there with them. It seems that some of the western and southern species which have been breeding in the Ohio River Valley or the southern Great Plains sort of drift eastward and congregate on some of these offshore islands. Hawks, falcons, and geese will be flying overhead.

BAY BONANZA

Those two massive estuarine areas, the Delaware and Chesapeake Bays, provide marvelous birding spots throughout the fall. A business meeting in Philadelphia is all that is necessary for an excuse to take your binoculars. Independence Hall and the Liberty Bell aside, wander out to the Tinicum National Wildlife Refuge just a mile from the airport, for a fine selection of eastern fall migrants. Some will be passing through, others will stay for the winter. If you have time to spare, drive down to the Bombay Hook National Wildlife Refuge on the western shore of the Delaware River where it meets Delaware Bay. Members of the waterfowl tribe meet and greet there in NOVEMBER.

JULY towards AUGUST finds shorebirds moving southward in the marvelous Chesapeake Bay area. That watery finger separates the larger portion of Maryland from its eastern part, known as the Eastern Shore. By SEPTEMBER, Common Loons may be seen and raptors will be present through OCTOBER. With a northerly or northwesterly wind, Point Lookout State Park could turn into a fall migrant trap. NOVEMBER brings Tundra Swans and Snow Geese in considerable numbers for this is a leading wintering ground for waterfowl. Try any of the many national wildlife refuges in the area.

Farther down the coast, the Outer Banks swing around mainland North Carolina catching a supply of fall migrants and wintering birds. Just don't head down there when gale warnings are up since birds won't be flying and ferries won't be running to the mainland. Scout out the beaches, the Bodey Island Lighthouse, Boston Woods in Hatteras, Pea Island, and then take the ferries to Ocrakoke and thence to Morehead City on the mainland. John Thompson, longtime birder, says not to miss the Sanitary Fish Market, a good restaurant.

John and his daughter, wildlife artist Lydia C. Thompson, live on Jekyll Island, off the coast of Georgia not far from the Florida state line. They think OCTOBER is the best time of year there. Much of the Georgia coast is lined with wildlife refuges, known generally as the Savannah refuges, fine for observing migrating species although they are most famous as wintering waterholes for waterfowl.

MIDCOUNTRY FLYWAYS

The midcountry flyways, the Mississippi and the Central, that migrating birds traverse by the millions in the fall, provide topnotch birding depending on the wind and the wet. Fall migration here as elsewhere begins in late JULY for some species. Shorebirds, waterfowl, and many passerines will be flying from Alaska and arctic Canada through the midwest down the great midcountry flyways. By AUGUST, shorebirds are beginning to move through Illinois and adjacent areas south of the Great Lakes. In the fall of 1987, American Birds reported 20,000 Pectoral Sandpipers at the Rice Lake Conservation Area in Illinois, while 257 Buff-breasted Sandpipers were congregating in the Riverton Water

Management Area in Iowa. Some shorebirds will still be found in appropriate habitats through *OCTOBER* and into *NOVEMBER*.

We think of Minnesota when we think of wild rice—and so do the birds. Migrating waterfowl feed on the wild rice and wild celery beds in *SEPTEMBER* and *OCTOBER*. Watch for the ducks and geese at Rice Lake National Wildlife Refuge, and be sure to bring home a package of the rice. Watch for Ross' Geese and European Wigeon in the more westerly wildlife refuges in the Great Lakes area.

HAWKING POLAR BEARS

OCTOBER's third week, or thereabouts, is the optimal time to scoot up to Duluth to catch the hawk migration. Wrap your scope in your long johns, for the northwest wind that brings in Northern Goshawks, buteos, and eagles can be chilly. Aim for the Hawk Ridge Nature Preserve. The colder it is, the better chance of seeing an early Gyrfalcon, Snowy Owl, Northern Hawk-Owl, Northern Saw-whet Owl, and just possibly a Great Gray Owl in the boggy boreal forests nearby. Migrating scoters, gulls, Bohemian Waxwings, Northern Shrike, Snow Buntings, Pine Grosbeaks, crossbills, and redpolls should be about. Resident grouse, Spruce, Sharp-tailed, and Ruffed are possibilities.

If between *OCTOBER* and mid-*NOVEMBER* you have the time and inclination, treat yourself to another of the world's wildlife spectacles, the

Polar bears are Churchill's star attraction in the fall

polar bears. Join a photo safari and fly up to Churchill. Watch the bears gather on the shores of Hudson Bay waiting for the ice to provide firm footing for their trek out to seal territory for the winter. Photographing the bears at close range is excitement enough, but the birder need not leave his binoculars home. With luck, you might behold a Snowy Owl or even a Gyrfalcon. Each afternoon when the tidal bay moved towards our polar bear camp several miles from town, a small flock of Purple Sandpipers flew up and back, stopping to pick out supper from the kelp. It was a life bird for most of us.

Waterfowl are featured in North Dakota from early fall until the arctic cold moves down and seeds ice on lakes and waterways. In autumn, waterfowl are counted in the hundreds of thousands. For an unusual travel taste, you could plan a feast of birding adventures in this state without fear of running into tourist jams. Visit each of North Dakota's 18 different wildlife sanctuaries and you will be surfeited on winged wonders. Try the Upper Souris National Wildlife Refuge for migrating waterfowl, five species of grebe, White Pelicans, Sandhill Cranes, Franklin's Gulls, migrant sparrows, and resident Sharp-tailed Grouse and Gray Partridge. Audubon National Wildlife Refuge hosts similar species; watch for whoopers. In SEPTEMBER, 100,000 Franklin's and Ring-billed Gulls roost at Tewaukon National Wildlife Refuge, and in OCTOBER and early NOVEMBER, 100,000 Snow Geese at Lake Alice National Wildlife Refuge. Throw some darts at the map of North Dakota and you will hit a good birding spot.

Midcountry is a big, big area. Some of the same species seen along easterly routes stream down through British Columbia and Alberta and on through the northwestern states. Schedule a fall trip to this magnificent region. In this part of the country, fall is pretty standard: SEPTEMBER to NOVEMBER. Sandhill Cranes are moving south in mid-OCTOBER over eastern Wyoming and Colorado on their way down to Bosque del Apache National Wildlife Refuge in New Mexico. Raptors will be moving through although not in the numbers as will be seen at eastern hawk watches. Redheads, Canvasbacks, Gadwall, Cinnamon Teal, and other waterfowl pass through. Although the fall waterfowl migration has been a major feature of the midcountry flyways for countless generations, it has in recently years dwindled alarmingly. The U.S. Fish and Wildlife Service estimated that the southern flight in 1988 was the second smallest on record, and that the duck population in North America has diminished by one-third in the past 15 years. Though indispensible to waterfowl, ponds and marshes are not profitable to farmers and industrialists. Hopefully, joint efforts between Canadian and United States wildlife officials and conservation groups will be successful in stemming the tide of wetlands drainage.

PLYING THE PACIFIC

Just as the Atlantic coast is favored in the fall, so is the Pacific. Beginning in late *AUGUST* and lasting through early *NOVEMBER,* fall is fun birding on the Oregon/Washington coast. Nice to have superb birding *and* spectacular scenery. Mount Rainier, shrouded in cloud and mist so much of the year, seemingly sits suspended in the sky on the cool, clear fall days. If you must choose the best time to visit this part of the country, try the middle two weeks in *SEPTEMBER.* It's a big area, with a fine variety of habitat, so don't rush your trip. Fly into Seattle or Portland depending on which state most attracts you.

SEPTEMBER is a peak month for shorebirds in many areas although anytime from mid-*AUGUST* through *OCTOBER* is good. Along northwest shorelines, expect a score of shorebirds including Lesser Golden-Plovers, Stilt Sandpipers, Surfbirds, Wandering Tattlers, the often sought but seldom seen Ruff, and Sharp-tailed and Buff-breasted Sandpipers. Looking for the Sharp-tailed? Try the Reifel Bird Sanctuary in Vancouver, British Columbia. If you are an ardent and knowledgeable birder, or if you go with an organized group with ardent and knowledgeable leaders, you will have some hundred-species days. When you are looking for a particular species at a particular time of year, it's a good idea to do your homework first.

SEPTEMBER and *OCTOBER* at the Malheur National Wildlife Refuge in Oregon, one of Roger Tory Peterson's dozen favorite bird-bashing places, should muster American White Pelicans, Trumpeter Swans, a "greeble" of grebes, and land birds ranging from Sage Thrasher and Sparrow, to MacGillivray's, Townsend's, and other western warblers. Sophisticated birders are always on the lookout for real rarities. The VENT bird tour organization recorded a Red-throated Pipit, a Siberian stray, in 1979, and a Temminck's Stint, a kind of sandpiper, in 1981.

Move inland from the ocean and shore to view a wide variety of habitat. Wet forests of the Cascade Mountains surely will produce Steller's and Gray Jays, plus Clark's Nutcrackers. That's the Clark of Lewis & Clark. Lewis got his name on a stunning pink-breasted woodpecker with an iridescent greenish black back. Watch for it, too, on a trip to this area. Alpine forest areas will give up Townsend's Solitaire and Warbler, Williamson's Sapsucker, crossbills, and maybe one of the three-toed woodpeckers. Incidentally, the Townsend of birding fame was an early 19th century ornithologist for whom Audubon named the Solitaire. Williamson didn't know beans about birds, but was an Army lieutenant in charge of the Pacific Railroad Survey who was honored by having the woodpecker named for him. Pothole country, riverine areas, isolated sagebrush areas, subalpine

Gray Jays are common residents in western mountain forests

meadows, and tundra grasslands are all productive of their special species. On an *October* trip, try Turnbull National Wildlife Refuge near Spokane for the annual fall migration spectacle.

Just about the same time, hawk-eyed birders will be at Pt. Diablo near the Golden Gate Bridge while they visit friends in San Francisco. Numbers aren't nearly as dramatic as in the eastern states, but late *September* through *October* normally produces good flights of Sharp-shinned and Red-tailed Hawks. Auk-eyed birders will be scheduling pelagic trips from Washington down the coast to San Diego. Land-locked birders will see all manner of migrating species from coastal viewpoints.

Anytime of year is good in California, but fall is the time that many experts choose. It begins early but continues to late *October*. Some shorebirds stick around through the winter. *August* to mid-*October* is best for landbirds too. Arnold Small, author of *The Birds of California* (and of a not yet completed second book about birds in his home state to be called *The Birdlife of California*) identifies the California coastal area as a focal point for fall migrants including some of the western warblers. When you tire of watching shorebirds, move up to the middle elevations of the Sierra Nevadas. Small advises prowling the perimeters of wet mountain meadows at the 6000 to 7000 foot level. Watch down in the coastal woodlands and especially on wooded promontories for Townsend's Warbler.

Most fall migrating birds are flying south, but some oddball birds insist on flying north from Mexico to the Salton Sea after their breeding season. It's known as the postbreeding reverse migration. Wood Storks, Laughing Gulls, and the Yellow-footed Gull (once considered a subspecies of the Western Gull) up from Baja are part of that reverse migration. If you can stand the heat of Hades, seek the Salton Sea. Collectors of Southern California endemics watch for the desert race of the Black-tailed Gnatcatcher. Perhaps it has already been designated a full species. Water Pipits abound there from early OCTOBER on. Western Grebes, nosing quietly around reedy waterways, will be back from northern breeding grounds. The high point of your fall trip could be a jaeger over this inland lake.

Offshore islands such as the Farallons are places to observe migrating seabirds. Well-known birdy spots such as Point Reyes National Seashore, Pacific Grove, Carmel, Santa Barbara, Point Fermin in Los Angeles County, Point Loma, and Imperial Beach in San Diego County are other special migratory spots. Coastal areas are also attractions for rare varieties and vagrants blown in on a high wind. Sharp eyes spotted Sharp-tailed Sandpipers and Ruffs in SEPTEMBER 1986.

In early OCTOBER, try Coastal California for Black Oystercatcher and Turnstone, Wandering Tattler, Surfbird, and Heermann's, Mew, and Thayer's Gulls. Monterey is a fine place to combine a couple of pelagic trips with some nearby land birding from Moss Landing down to Point Pinos. Move farther south to the Santa Barbara area for the same combination. Fall migrants will still be going through and winter visitors will have arrived.

PELAGIC PERHAPS?

Pelagic trips offshore from Washington down to southern California are in full swing in the fall. Migrating seabirds begin moving south from the high Arctic in AUGUST. Fear not that you have never strayed from *terra firma* before. It's easy and it's fun. A respectable North American bird list is not really "respectable" without at least a few pelagic species. As is true with any pelagic trip, it's best to schedule more the one per visit although California trips are rarely canceled because of weather. Many trips are run from AUGUST through NOVEMBER providing a better composite experience if you try more than one oceanic location.

Peak pelagic season is AUGUST through mid-NOVEMBER. Shearwater Journeys runs a lot of reasonably priced pelagic birding trips off California and is an ideal choice for novice and expert alike. Debi Shearwater is both enthusiastic and knowledgeable about the birds of the sea and about the whales and porpoises, too. Debi will quickly orient you to the most important factor leading to success other than remembering to bring your seasick pills. "We call out the location of the birds by the o'clocks," she will tell you in the first three minutes of her orientation held about 6 A.M. Noon is "12 o'clock, dead ahead" (no matter which way *you* are facing). The o'clocks are standard birding terminology and absolutely essential on a boat.

Listen and learn quickly so that you waste no time looking across the generally featureless water as someone calls out, "Long-tailed Jaeger, 3 o'clock halfway between the boat and the horizon." There, you saw it! A lifer. Pink-footed, Buller's and Sooty Shearwaters are certainties on most trips, but your birding companions will be watching intently for those errant rare species. In *AUGUST* 1986, it was a Wedge-tailed Shearwater, believed to be the first report of the species in North America.

In California, trips go out from San Diego, Santa Barbara, Monterey, and San Francisco. A San Diego pelagic trip would make a boatload of birders pleased with Craveri's Murrelet, to say nothing of Least and Black Storm-Petrels. They would be ecstatic with a Red-billed Tropicbird. You won't have a hundred-species day out on the ocean, and if you have been out several times before you may only see one new species—or maybe none.

Three successive California trips one *AUGUST* put two tubenoses, members of the large family Procellariidae, on my list. Those are the birds that have a tube, or nostril, on their bill which enables them to excrete the salt from seawater. My two were shearwaters, Buller's and Black-vented. I had new views of Pink-footed and became more familiar with Sooty Shearwaters, the latter floating on the sea in great rafts. I also added to my list Leach's and Ashy Storm-Petrels, Cassin's Auklet, and the third, and my last, species of jaeger, the Parasitic. Highlight on that trip for many of us was a South Polar Skua, and for me a lovely Sabine's Gull. Any pelagic trip will produce some highlight, from a new species of bird, to a new whale, to a new friend.

OCTOBER pelagic trips along the Washington and Oregon coast are justifiably popular. Characteristic sightings are Black-footed Albatross, Buller's and Pink-footed Shearwaters, Fork-tailed Storm-Petrels, all three jaegers, Cassin's Auklet, and a good selection of Alcids: Rhinoceros Auklets, Tufted Puffins, Marbled Murrelet, and Pigeon Guillemot. Arctic Terns will be at the early stages of that longest migration trek of all as they leave their breeding grounds in Alaska and the high Arctic to seek "summer" in Antarctica. The South Polar Skua does just the opposite. It

Birder checks field guide during dull moments

"winters" in the Arctic during our summer and breeds in the Antarctic austral summer during our winter. Red and Red-necked Phalaropes should be seen but they are hard to tell apart at this time of year. Closer to shore, you should see Heermann's, Sabine's, and Mew Gulls, Harlequin Ducks, Brandt's, and Pelagic Cormorants.

Pelagic birding can provide some of the most exciting birding of all, but much of the time out on the endless ocean, it can be dull and boring. That's the time to glance furtively at your bird guide to find out what everyone else is talking about when they were told there was a chance of seeing *Pterodromas.* You didn't want to admit you had never heard of them. Quickly you discover that they are large petrels, often called Gadfly Petrels. The boat will not need to go far out for you to get a glimpse of members of this genus, a Black-capped in the Atlantic or possibly a Mottled or Cook's Petrel in the Pacific.

SOUTHERN LIVING

If warmer climes are more up your birding alley, go back to Texas. Not only is it the place to be in the spring, it's super in the fall, too. Victor Emanuel, whose birding tour organization (VENT) bears his name, is one of the foremost experts on the birds of Texas. He suggests, "OCTOBER is probably the month to look for Mexican vagrants and rarities." He's talking about such species as the Northern Jacana and about the Masked Duck, closely related to the Ruddy Duck. Even if you miss them, you will be pleased to see the Mexican Crow, Buff-bellied Hummingbird, Great Kiskadee, and Ringed Kingfisher. You may not see them all, but you might see some specialties from south of the border such as Red-crowned Parrots and Green Parakeets.

Great numbers of water and land species wing from the north down to the southern tip of Texas at this time of year. Additionally, south Texas is famous for resident specialties such as Green Jay, Altamira and Audubon's Orioles, and Common Pauraque. Grooved-billed Anis are getting ready to head south as is Couch's Kingbird. Generally the area between Harlingen, south to Brownsville, and west along the Rio Grande River to Falcon Dam beyond Rio Grande City is the best. Be sure to visit two favorites: Bentsen State Park and nearby Santa Ana National Wildlife Refuge.

Up in Oklahoma, there's a fall gull fiesta with a reputed three million Franklin's Gulls stopping to rest and feed on their way to wintering sites along the western coast of South America. Try the Washita National Wildlife Refuge. In early OCTOBER, Sandhill Cranes may stop here on their way to Aransas. In NOVEMBER and DECEMBER, lots of ducks and geese stop over.

Fall birding along the Gulf Coast can be good as the migrants gather along the beaches preparatory to flights across the water to Central and South America. Birding writer Judith A. Toups recommends the tiny coastal area of her home state, Mississippi, from late August to late October, depending on when a tropical storm or hurricane crashes in. Down in Mexico, try your hand at tropical bird identification. Head to Manzanillo on the Pacific coast and to nearby Colima. Experts will tell you that this area of beautiful scenery is one of the most productive areas, one to stimulate a taste for the neotropics and neotropical birding. A phenomena you will learn is the tremendous variety of birds in this part of the world. Quickly you will add a hundred birds to your life list. (You may be smart not to keep a list!) You will recognize familiar warblers, flycatchers, vireos, and finches, in addition to trogons and parrots that you may have seen in Arizona and Texas. You may not yet be conversant with bird names like euphonia, grassquit, saltator, and attila but they will pique your fancy. Discover also the hyphenated birds: silky-flycatchers, brush-finches, magpie-jays, ground-cuckoos, wood-rails, and ant-tanagers, to mention a few.

MONARCH'S TIME

While kingfishers, kingbirds, and their peripatetic feathered friends progress southward from late summer until early winter, another kind of king, the Monarch butterfly, attracts our attention nearly everywhere. We watch the gentle movement of these flimsy creatures as they make their miraculous journey of a lifetime. Whether it's at Monarch Pass in the Smoky Mountains where they stream through in astounding numbers, at the Cape May Hawk Watch, or along the west coast, we marvel at the force of life that guides such beautiful and delicate creatures. Upwards of half a billion monarchs make their way over thousands of miles from as far north as Canada down to their winter roosts in the mountainous fir forests of central Mexico. It is a truly royal feat.

RUN FOR THE RARITIES

Waters offshore from both coasts invariably turn up species to set the birder's heart palpitating: Black-capped Petrel off North Carolina, Least Storm-Petrel off California in *SEPTEMBER* 1987, Short-tailed Shearwater and Craveri's Murrelet off California in *OCTOBER,* and a Little Curlew in a California field in *SEPTEMBER* a year later. In *OCTOBER* 1987, Yellow Rails were seen in Tennessee, Louisiana, and Texas, and Ontario hosted Lesser Black-backed Gull, Little Gull, and Northern Sawwhet Owl. Are these on *your* list? Eurasian Wigeon showed up in small numbers all over the place.

Other than the susceptibility of coastal areas to rare species, Texas probably holds the record. The big state closed out the 1988 winter season with a bang in *JANUARY* with well-attended sightings of a stunning Crane Hawk in what was believed to be its first visit of this Latin American species to North America. Other winter wonders from south of the border included Hook-billed Kite, Tropical Parula, Graycrowned Yellowthroat, Golden-crowned Warbler, Crimson collared Grosbeak, Blue Bunting and White-collared Seed-eater. Most of these are ho-hum species south of the border, but purist North Americans get all excited when they see them in Texas.

Six

Winter:
A Time for Resting

Winter provides the testing months, the time of fortitude and courage. For innumerable seeds and insect eggs, this period of cold is essential to sprouting or hatching. For trees, winter is a time of rest. It is also a season of hope. The days are lengthening. The sun is returning. The whole year is beginning. All nature, with bud and seed and egg, looks forward with optimism.

Edwin Way Teale
Wandering Through Winter

*W*inter, a gray and dismal time for many people in the northern hemisphere, speaks of cheerless cold, stinging winds and pale sunshine on glistening snow. Yet winter in the lower latitudes of this hemisphere is bright of sky, painted with colorful flowers, and warmed by softly whispering breezes. To the snowbound, winter in the southern states sings a siren's song. Winter, like other seasons, has its special times and places for the birder. Listen to the hoot of the Great Gray Owl where howling winds do blow. Listen to the noncommittal "chip" of the wintering warbler where palm leaves rustle.

Clark's Nutcracker's are noisy birds found at timberline in western mountains

Though it lacks the restless movement of migrating species, and the feverish feeding routines of summer nesting time, winter (barring blizzards) often is a good time for birding. Mountain birds like Clark's Nutcracker and Gray Jay are more easily seen at lower elevations. Stately wading birds are easily observed from boardwalks in southern wetlands. It is a time of birding spectacles as great masses of waterbirds gather in winter watering grounds, or as eagles gather where the salmon run.

WHEN IS WINTER?

This last season of the birding year runs more or less from *DECEMBER* until the end of *FEBRUARY.* Because most wintering species are stationary at this time of year, precise timing for special observations is not as important in this season as it is during the migratory seasons. True, in the high Arctic, winter may begin in *SEPTEMBER* or *OCTOBER,* and it's not uncommon for snow to fall high in the Colorado Rockies in *SEPTEMBER.* In subtropical south Florida, traditional evidences of winter—snow and ice—just never appear. (Well, once it did get really cold and some people reported a few snow flakes.)

By late *FEBRUARY* winter begins to usher in the new spring of the birding year. Spring begins then to move northward up the continent before winter has left the thoughts of most North Americans. Northern Cardinals begin to sing their territorial songs though snow and ice still hide the berries and grasses that will feed a world of animals in the summer. Like other seasons, winter begins at different times at different latitudes and altitudes. Winter, perhaps, is a state of mind.

Winter birding groups tend to look south, to south Florida, southeast Arizona, southern Texas, and farther south to Mexico and South America. For the hardy birder who would sample the whole calendar range, one can find winter trips to Newfoundland, Duluth, Ontario, and other brutally cold places where select species may be seen on their favorite turf.

CHRISTMAS COUNTING

As official winter begins, the greatest event of the North American birding year takes place: the *Christmas Bird Count.* Often referred to as the CBC, or even just *the Count,* it is sponsored annually by the National Audubon Society throughout North America. Counts beyond continental borders have been increasing and, in 1987, included several locations in Mexico, Costa Rica, Panama, Belize, Trinidad, Tobago, Hawaii, Guam, Saipan, Bahama Islands, Dominican Republic, Puerto Rico, Virgin Islands, and Bermuda.

Although no one can accurately estimate the number of birders there are in North America, we do know precisely how many are eager enough to crowd a full day of birding into an already crowded holiday schedule. In 1987–1988, it was 41,920: 34,865 field observers and 7055 feeder watchers participating in 1544 separate counts. Birders swatted mosquitoes in the Everglades National Park and shivered in the snow in Fairbanks Alaska, but they were doing it because it was a combination of fun, adventure, and companionship. Not only that, this vast volunteer effort supplies important knowledge to the scientific world.

According to Susan Roney Drennan, editor of *American Birds,* the Audubon ornithological journal that publishes an annual compilation of results, the Christmas Bird Count is "the single most popular, voluntary, early winter bird continental inventory in the world." She was writing about the 85th annual count in 1984, calling it "the most extensive, longest-term, continuous, and most geographically comprehensive data set in American ornithology." Wow!

Although the spark of enthusiasm for the CBC varies in different parts of the country and within different Audubon chapters, the number of actual participants is but a fraction of the populace who call themselves bird watchers or birders. One might think that numbers of participants would also have something to do with climate—the more salubrious, the more counters. Not necessarily so. In spite of the fact that Florida is the winter home for many bird species and that the state's population mushrooms in the winter, not one of the Florida area counts in 1987–1988 had more than 100 participants. *American Birds* reports on the 88th count that such northern places as Victoria and Vancouver in British Columbia; Edmonton, Alberta; Cuyahoga Falls, Ohio; and Ithaca, New York, all had over 100 counters that year and Hartford, Connecticut, topped the list with 153.

Often it is the dogged dedication to our task that shines brightest. Count day in Prudhoe Bay, Alaska, not only was an overcast, minus-temperature day, but the count must have been discouraging to Edward Burroughs who provided the official figure: three Common Ravens, period. The Fairbanks Bird Club did better. Rousting out of bed in the early darkness, 27 counters birded 356 miles on foot, by car, on skis, and by dogsled to list 26 species with Common Redpolls getting honors for highest number of individuals. At the other end of the weather spectrum, hordes of mosquitoes in some parts of the Everglades National Park would carry off the intrepid bird counters were they not wearing

Miami Christmas Bird Counts can be pleasantly warm and relaxing
(Courtesy Jack Holmes)

mosquito hats. Ever tried using your binoculars through the netting?
Checkerboard birds. Closer to downtown Miami, the backside of a museum provided a pleasant setting for a relaxing count of birds passing
through the canopy, and no mosquitoes.

No matter where you are during the winter, opportunities to see
birds abound. You might think that the action at Sun Valley, Idaho,
would be confined to the slopes. Not so. CBCers found 46 species of
birds in 1987: lots of Mallards, Black-billed Magpies, Rock Doves, European Starlings, and House Sparrows. In all, they added up to over
4000 individual birds. However, if you had been with tour leading
guru Victor Emanuel and 146 other birders on the Freeport, Texas,
count, you would have exulted in a record 215 species. South Texas
clearly is a good place to bird in winter. The king-sized state attracts
top numbers of species and top numbers of birders, though California
certainly seems to have put the top number of counters in the field. If
you have never been on a Christmas Bird Count, try it next year. You
will meet the best birders in your area, perhaps in the country, and
have the satisfaction of participating in a significant national event.
Most of all, you have fun.

SNOWBIRDING IN THE SUNBELT

All across the Sunbelt, northern nesters are resting. From southern Florida right across the continent to southern California and down to Baja, you will find good, even spectacular birding. Northern waterfowl will be concentrated in massive flocks on southern lakes, ponds, and waterways in coastal areas. Shorebirds, so confusing in their winter plumage, linger like gray blankets draping southern coastal beaches.

DECEMBER through *FEBRUARY* on birder's winter calendars is prime time for Sunbelt birding. Showbirds—herons, egrets, ibis, storks—are easily seen along roads, trails, and boardwalks through the Everglades National Park, the wildlife drive through the Ding Darling National Wildlife Refuge on Sanibel Island, or in other favored habitats. It's the time of year when the Everglades, for the most part, are bearable rather than buggy. American White Pelicans choose south Florida for a winter holiday and birders will see grand formations of great white birds floating on an air cushion low over Florida Bay.

South Florida visitors driving down through the Everglades National Park, pass through a unique part of North America and of the world. Thanks to Marjorie Stoneman Douglas and her book, *River of Grass,* widespread attention has been given to preservation of the Everglades. Many a visitor has been turned on to birds while walking

Little blue Heron shares Corkscrew Swamp boardwalk with birders

the park's numerous boardwalks and paths. Egrets, ibis, herons, and bitterns share the paths with birders and zealous photographers. Out on the islands in Florida Bay, Roseate Spoonbills cluster in large nesting colonies. Follow instructions from locals for views of Greater Flamingos.

Waterbirds, waders, raptors, gulls and terns, woodpeckers, and warblers are easily seen. You don't even have to try very hard to see 60 to 70 species in a day. South Florida is a good place in winter to see both the northern and southern races of the Bald Eagle. Late in the winter, watch for the graceful Swallow-tailed Kite, a beautiful sight in anyone's day. The little feathered rainbow, better known as the Painted Bunting, is a regular visitor, particularly at Castellow Hammock Park in South Dade County.

Experienced birders will seldom see a record bird in this environment, but early and intermediate birders will have good days at nearly any park or wildlife sanctuary south of midstate, and records are always possible. In 1987, a Black-faced Grassquit from the Caribbean made one of its infrequent forays into Florida. The small finch-type bird was seen for several weeks in a brushy dump near the Broward County Airport. A couple of years later, it was across the water from that county's seaport that birders from all over the country came in early February to admire the Bananaquit, a Honeycreeper, on one of its infrequent visits from the West Indies.

You may be startled by the sight of Sandhill Cranes lifting off gracefully from along side of the road. They are good "car birds," quickly and easily identified as you drive in the winter in southern Florida. You can even see these imposing birds as you drive along the turnpike. Keep watching. Peer through the marshes at the Loxahatchee National Wildlife Refuge near Palm Beach to see the fine concentration of Fulvous Whistling-Ducks most years. Pretend to be an Indian while you canoe along winding water trails. In the Audubon Corkscrew Swamp Sanctuary, endangered Wood Storks gather. Scouting feeding grounds from high up in the puffy clouds, these grand black and white creatures circle effortlessly across south Florida skies the better part of winter.

Birders driving RV's, as many do, enjoy birding the Sunbelt, generally finding some of the best birding along the Gulf Coast. A necklace of wildlife sanctuaries adorns the coastal areas from Florida to Texas. A stop at St. Marks National Wildlife Refuge where Florida bends around the Gulf of Mexico is a good idea. Check off a good collection of eastern birds in this refuge with its salt and brackish marshes, hardwood swamps, pinelands, and oak forests. Waders, wintering waterfowl, woodpeckers, and wrens reflect the variety of habitat that encourages the wintering species you would expect to see.

Early in the winter, you will likely see Bald Eagles nesting or watch them catching supper. A hapless grebe one evening provided a pair of eagles the opportunity for careful calculation of just the right moment to swoop for supper. Taking turns from perches on opposite sides of a small pond, they made their passes. During the first few dive-bombing attacks, the grebe successfully eluded capture. But alas, it surfaced when it should have ducked. Gourmet grebe on the menu that evening. Gourmet birders will prefer the splendid Apalachicola Bay oysters.

JANUARY will be the peak month for waterfowl in most of the wildlife refuges throughout the South. Where woodland and wetlands are combined, as they are in many refuges, woodpeckers, warblers, and other small species will provide good winter birding. Wood Ducks will be nesting in many refuges but other ducks begin their migration north by *FEBRUARY.* Coastal Louisiana, the southern terminus of the Mississippi flyway, is alive with wintering ducks and geese. South of New Orleans, Delta National Wildlife Refuge provides haven for upwards of a quarter million birds. It's a watery wilderness, so expect to inquire about boating trips.

WHOOPING IT UP

Mid-*DECEMBER* signals the arrival in Texas of that state's most famous avian winter visitor, the rare and endangered Whooping Crane. This majestic bird, whose retreat from extinction has brought cheers from naturalists throughout the world, is one of America's two species of cranes. Each year, these wonderfully dignified birds travel between their summering grounds at the Wood Buffalo National Park in Canada's Northwest Territories and their wintering grounds at the Aransas National Wildlife Refuge, a 2500 mile trip. A boat trip out of Rockport provides the best opportunity of seeing the cranes. The "Whooping Crane" boat that plied the inland waterway to the avian Whooping Crane has been supplanted by Captain Ted Appel's *Skimmer* designed with the needs of birders in mind. The three-hour trip that leaves from the Sandollar Pavillion, goes across Aransas Bay and out to Matagorda Island, and back again, has turned into a real birding adventure. Your captain even provides a bird list, carries a field guide, and has a scope mounted on the upper deck rail. You may be lucky and see the cranes close to the waterway or they may just be white dots on the distant green marsh. A scope will be helpful as the boat frequently beaches along the calm waterway and the viewing deck is stabilized. Captain Ted takes the *Skimmer* on spring nesting jaunts too.

All up and down the Texas Gulf Coast from Louisiana to the tip

Whooping Cranes are seen at a distance from boats

end of the state, a string of National Wildlife Refuge's furnish safe resting areas in coastal marshes for all manner of waterfowl. Refuge-hop or throw a dart; you will see wintering wildlife by the tens, maybe hundreds of thousands. Greater White-fronted and Snow Geese reach peak numbers in *DECEMBER* and *JANUARY*. Look carefully at all those white geese for you may spot the rare Ross' Goose.

A typical *DECEMBER* south Texas birding tour can reward partici-pants with well over 150 species. Go with a friend or join an organized trip and meet new friends. South Texas is a good time and place for birders at every level of expertise. Those who just want to see a lot of species derive just as much satisfaction as those twitchers who run for the rarities. Mexican vagrants

Wintering migrants in sunny south Texas will be side-by-side with a few from south of the border. Regular Mexican visitors may include Clay-colored Robin, Roadside Hawk, Hook-billed Kite, Golden-crowned Warbler, Ruddy Ground-Dove, or Crimson-collared Grosbeak. Try for Red-crowned Parrots and Green Parakeets in Brownsville and McAllen residential areas. Now how about telling your Minnesota birding friends about those birds!

In Brownsville, be sure to visit one of the most famous of birding meccas, the city dump. Here you will encounter friendly garbagemen,

Brownsville, Texas, garbage dump is famous birding area

other birders, and the object of the trashy quest, the Mexican Crow. This small, glossy crow, belonging on the other side of the Rio Grande River, can be found fairly regularly at the dump. On a recent visit, I encountered a birder from Vermont and two birders from Finland. They added both the crow and the Chihuahuan Raven to their life lists. This dump, like most others, attracts scavenging gulls. Most are the common garden variety, but every so often some sharp eye will spot a Lesser Black-backed, Thayer's, or Glaucous Gull. For birders with an acute sense of the ridiculous and the sublime, top off your dump experience with lunch at the elegant Palm Court Restaurant. You will be provided a fine menu and tantalizing desserts such as one called "better than sex." Though your scavenging attire may add a jarring note to dining beneath a crystal chandelier, the owners have learned that their dump is famous and you will be welcomed with gracious good nature.

Head then for the Santa Ana National Wildlife Refuge and Bentsen State Park. At Bentsen, campers regularly put out hummingbird feeders, oranges to attract Altimira Orioles, and all manner of scraps favored by noisy Chacalacas (with a name like that, you can imagine how a dozen of them sound all at once!). Both areas are good for Mexican visitors like the Tropical Parula and recently the Gray-crowned Yellowthroat. Like south Florida, south Texas provides a salubrious climate attracting a variety of regular visitors, vagrants, rarities, and exotics. A strident call overhead on a *DECEMBER* 1988 visit to Bentsen, brought to binocular-view a beautiful white parrot with sulphury yellow underwing feathers. This birder recognized the Little Corella, a species native to the Darwin area in Australia. Even an "uncountable" bird can be exciting.

Falcon Dam, farther up the Rio Grande, is another birding area favored among birders. Star of the show is the Ferruginous Pygmy-Owl found downstream near the site of the old Girl Scout camp. It won't pop out and say "Howdy," but with patience, and either good guidance or good direction, you may find it. Playing a tape may help, but it may just call up another birder playing a tape in answer to your tape. How embarrassed can a tour leader be! Watch down at riverside for the Green and Ringed Kingfishers, and overhead for the tropical Gray Hawk. A nearby campground is a good place to show a first-time Texas visitor the cocky Greater Roadrunner getting a free breakfast, and lunch, too.

Coastal and southern Texas is not the only good wintering area for sought after sightings. Muleshoe National Wildlife Refuge northwest of Lubbock, the oldest such refuge in Texas, boasts the largest winter concentration of the Lesser race of Sandhill Cranes in the United States. Numbers peaked in *FEBRUARY* 1981 at more than a quarter million. The cranes have plenty of company there for ducks like the lakes of the high plains, too. Some are winter visitors; others nest there.

With time on your hands and waterfowl on your mind, you might drive north of Dallas to Oklahoma and visit the Tishomingo National Wildlife Refuge just across the Texas border. Large numbers of ducks and geese that came down the Central flyway are resting there. Peak time is between *NOVEMBER* and early *FEBRUARY* when they begin winging north. Located just east of the 100th meridian, it is a meeting point of eastern and western passerines.

Winter birding in the southwestern United States is superb. Mid-*JANUARY* is the time selected by the experts, but most any time during the winter will be a good time. The 1987–1988 Christmas Bird Count east of Phoenix counted a very respectable 152 species, while the San Diego Count racked up a smashing 200 species.

New Mexico and Arizona offer fine winter birding. The lower Rio Grande River valley is a good focal point if you are visiting New Mexico to see western species, wintering waterfowl, and raptors, as well as a few rarities. Check your map for the many National Wildlife Refuges and visit as many as time allows.

Top of the list is the famed Bosque del Apache National Wildlife Refuge. In addition to the cranes, Ross' Geese ("want" list of many a birder) can be seen in small numbers. It is at Bosque that many thousands of Snow Geese and Sandhill Cranes may winter over. The Sandhills at this refuge have been in the foster parent program. They have been apparent willing participants in an experimental program involving brooding the eggs and bringing up some of the very rare Whooping Cranes. The refuge's riparian habitat along the Rio Grande River offers a salubrious setting for many native and wintering birds. Scrub Jays, Cactus, Rock, Canyon, Bewick's, and Marsh Wrens should make a dent in your wren list. Watch for Curve-billed and Crissal Thrashers, sparrows, juncos, and Lesser Goldfinches.

Border areas, are good for vagrants (not the ones who break into your house, but the avian variety). Look not for wetbacks, but for Streak-backed Oriole, Rufous-backed Robin, and Green-backed Heron in a reservoir somewhere. You might catch the glint of a rare Green King-fisher. Anna's Hummingbirds, Black-tailed Gnatcatchers, Townsend's Solitaires, and Lawrence's Goldfinches regularly winter here and there's a good chance of seeing Montezuma Quail. For genealogical buffs, Anna was a Duchess of Rivoli back in Napoleon's time, Townsend was an ornithologist friend of Audubon's who named, as well as painted, lots of birds, and Lawrence worked with other ornithologists of the early day when the true early birders were naming birds for each other.

In southeast Arizona, watch for Chestnut-collared Longspurs, Yellow-headed Blackbirds, Lark Sparrows along with the Buntings, and two sage (if not smart) birds, Sage Sparrows and Sage Thrashers, along with some of the other thrashers. For a tour of the southwest, Arizona and southern California make a good combination. Delightful weather, marvelous scenery, superb winter birding, and good air connections between blizzards of the north and the palms of Los Angeles or San Diego, are an unbeatable combination. The terrain offers just about every habitat designed to build up your birding list: seacoast, swamps, mountains, woodlands, and desert. Tick off the southwest natives and the wintering birds from points north, as well as the occasional rarity. Get off the interstate as often as possible to where you can pull off the road and scan fields, hills, and sky without fear of being smashed by an onslaught of trailer trucks.

Depending upon the habitat and location, the far southwest provides a good opportunity to hone your identification skills by separating out Pelagic and Brandt's Cormorants, or Western and Clark's Grebes. Build your list of black birds: Black-shouldered Kite, Black Turnstone, Black-tailed Gnatcatcher, Black-throated Sparrow, and the Tricolored Blackbird. Many birds are not easy to find, some very localized. Take your Lane guides along, go with an organized birding tour group (check with local Audubon groups and bird clubs), or just take your chances gypsy fashion.

With homework done, you should find Ross' Goose, Mountain Plover, Abert's Towhee, Strickland's Woodpeckers, and in mountain areas, Montezuma Quail. In woody meadows of California's San Joaquin Valley, watch for congregating Yellow-billed Magpies, seed-seeking sparrows, and a few hummingbirds. Thrashers thrash the southwest pretty well and in addition to the common Curve-billed Thrasher and the wintering Sage Thrasher, you just might find Crissal, Bendire's, or even a Le Conte's Thrasher. Not all habitats are pretty. A dusty, dingy old oil field near Maricopa, California, with patience, may produce a Le Conte's Thrasher. This Le Conte incidentally, was Dr. John Lawrence Le Conte, a renowned entomologist and cousin of the other Dr. John Le Conte who taught physics and chemistry, for whom the sparrow is named.

Be sure to visit the Salton Sea area, the famous migrant trap into which so many wintering birds happily fall. Guy McCaskie reported in *American Birds* that an astounding two and a half to three million Eared Grebes had been seen there on *JANUARY* 23, 1988. It is tempting to visualize what a sight that would be in summer plumage: a yellow sea of five million golden tufts bobbing on the water! Alas, the winter grebe is pretty drab. A little earlier, on the Christmas Bird Count, there were thousands of ducks and geese including 1006 Ross' Geese, 3038 Northern Pintail, and 8880 Northern Shovelers. Waterbirds are certainly not the only birds to be seen. Mountain Plovers can be found in substantial numbers and White-throated Swifts are nearby along with a catholic sampling of owls, hummingbirds, flycatchers, warblers, and sparrows for a respectable total count of 154 species. This count attracts some of the birding bigwigs, so why don't you join them next year?

Northern California in winter rivals the southern part of the state in tantalizing birding. During mild winters you might find Snowy Owls. Black-backed Woodpeckers and Bohemian Waxwings will stir up warm blood on a cold day. Coastal species could be Emperor Geese, and if you look carefully at the Green-winged Teal, you might spot the Eurasian version. The whole state of California is flush with birds and birders: thousands of birders saw hundreds of thousands of birds on the Christmas Bird Count. A record of 33 separate counting groups each recorded more than 150 species on that day for the official 1987 count.

TROPICAL DELIGHTS

If time is no problem on this ideal trip to the western end of the Sunbelt, or if you have always wondered about that long skinny peninsula south of California, go on down to Baja. For birders and whale patters, you can't beat it. Vermont birder Maida Maxham, escaping minus temperatures at home, eloquently described her first visit:

Baja California is a study in contrasts, desert surrounded by sea: tiny slow-growing cacti and succulents, extravagantly huge blue whales spouting, gray whales spyhopping, elephant seals bellowing, and of course birds. Heerman's gulls and yellow-footed gulls are never far from the breeding colonies of California Sea Lions or Northern Elephant Seals, ready to clean up the afterbirth during pupping.

As the desert comes into bloom, hummingbirds zoom from blossom to blossom, and the silent cactus wren investigates those strange plants for choice morsels. Mangroves shelter Marbled Godwits, curlews, and Black Turnstones on their winter holidays. The rocky cliffs are home to Osprey and their miraculous piled-up nests. Baja is a place to absorb the enormity of blue whales in their independent, remote grace, or drift in small boats among the friendly gray whales, interacting with them, looking eye to eye, petting them. Baja is also a place to swim with sea lions or watch a Peregrine Falcon, perched on a cliff, streak into the air after a Least Grebe and see it return stroking slowly back to that cliff with its trophy. The bioluminescence of the sea at night is pure magic. Maybe somewhere there are banditos.

Sounds like a place right out of a travel magazine, doesn't it. Well, it is. Many tour operators go there: check both birding and whale watching expeditions, or take your camper, binoculars, and snorkel gear.

Winter vacations in neighboring Mexico have long been favored by birders and tourists alike. Birders like Oaxaca for Christmas; Colima and Jalisco in *JANUARY*; and Palenque and the rest of the Mexican state of Chiapas, and the Yucatan in *FEBRUARY*. Actually, most any winter week spent birding in the vicinity of some of the world's greatest temples—Palenque or nearby Tikal in Guatemala—will bring new delights each day during long walks in and around the ruins. Jim Clements, author of the world birds checklist, suggests that Mexico may well be the best place for North American birders to get their first taste of foreign birding.

As late as *MARCH,* the combination of templeing, snorkeling, and birding can't be beat. Mayan temple ruins such as Chichen Itza, Coba, and Uxmal are all excellent tropical birding areas. Located in dense jungles, partially cleared away in rather recent times to reveal ancient splendors, the temple sites have been untouched for centuries. They offer a shining opportunity for the two-interest family: the birder spouse seeking sparkling feathered creatures in the air, and the fisher/snorkeler spouse who wants to explore Caribbean waters for colorful scaly creatures below the surface. For both, there is the special appeal of Mayan history through which you will be walking. Spend a week at ruins like these.

Do remember, tropical birds arise early in the morning. If you want to see them, follow their example—*every* day. There is nothing like the jolt to the foggy brain delivered by the sight of a turkey-sized Crested Guan greeting the earliest of morning sunrays from a low tree branch over the path in front of you. You may not ever see it again. Southern Mexico offers habitat variety from watery lowlands to high mountain ridges. You might even discover there is another roadrunner, the Lesser Roadrunner, running out of the swirling clouds at El Sumidero near Tuxtla. Winter is truly an exquisite time of year in Mexico, Central America, and the many nearby island paradises. The posters and travel magazine advertisements tell it true: colorful flowering lands, azure coastal sea vistas, and puffy white clouds. Birders enjoy the settings and the birds, too. The Bahamas, Jamaica, and Costa Rica are rich in birdlife throughout the winter and early spring. Just pick your paradise.

COLD "HOT SPOTS"

"I don't think I'm *really* birding unless I have to get started well before dawn, the weather is cold with blustery winds, the best place requires hiking along rough trails, and getting your feet cold and wet," Pennsylvanian Paul Beach said when we were talking about the delights of south Florida birding in the winter. Some people really like to punish themselves: hold a blanket around their shoulders with one hand and grasp frigid binoculars in the other.

There's plenty of that kind of birding during North American winters. Birders are a hardy lot. At this time of year they stare with single-minded intensity at the gaps in their life lists and head for the Parker River National Wildlife Refuge on Plum Island near the charming town of Newburyport north of Boston. They would be happy with a wintering Snowy Owl and maybe a Northern Shrike. Gull fanciers will be looking for an isolated Iceland Gull, the Common Black-headed Gull (note the capital *C*) recently arrived from the old country, or a Glaucous Gull.

Cape Ann, jutting out into the sea south of Newburyport, should muster Great Cormorants, Purple Sandpipers, Common Eider and Goldeneye, or possible a Barrow's Goldeneye. Harlequin Ducks and ocean birds such as Black Guillemot, Razorbill, or Dovekie may be out on that cold, gray deep. The Boston area has some of the most experienced and enthusiastic members of the birding brigade on the continent, some of whom saw 113 species on the 1986 CBC including 21 Snowy Owls.

Northern Minnesota around the western tip of Lake Superior, may be a good bet if you are lacking owls and looking for bracing weather. Names of the species likely to be seen in early *FEBRUARY* constantly remind you of where you are: *Northern* Goshawk, Hawk-Owl, and Shrike; *Boreal* Owl and Chickadee; *Spruce* Grouse, *Hoary*

Redpoll, *Pine* Grosbeaks, *Snow* Bunting, and *Snowy* Owl. Duluth is probably the best place in the United States to see the Snowy Owl. Everyone will be hoping for a Gray Jay on a gray day, and a Great Gray Owl to boot.

Field Guides, Inc., a bird tour organization led by some super-birders, likes Amherst and Wolfe Islands at the eastern Ontario shore in late *FEBRUARY* to early *MARCH.* Howling owling up there—Snowy, Short-eared, Great Horned, Long-eared, Eastern Screech, Saw-whet, and Northern Hawk-Owl—is on the menu. You just don't see many of them in sunny south Florida although Great Horned Owls are not uncommon and Eastern Screech Owls reside in a dense Sago Palm in my south Miami garden.

If you are still thinking cold, try Newfoundland with Stuart Tingley, a Wings, Inc. tour leader who lives in nearby New Brunswick. Actually, he says it's not as cold as it sounds due to the offshore Gulf Stream. In addition to a good variety of northern gulls, eiders, and alcids, especially be looking for some European immigrants for your North American list: Tufted Duck, Common Greenshank, Fieldfare, and Redwing (not Blackbird, just Redwing, a small European thrush).

At the far western reaches of the continent, be really adventurous and visit Kodiak in the Aleutian Islands. It may not be as cold there as

Eastern Screech Owl is easily seen in the eastern United States

you might think, the temperature in Kodiak for the Christmas Bird Count in 1986 was in the 26 to 34 degree range. They had a whopping good Count too: 76 species to capture Alaska's title for the largest number of species. You could add a few to your list up there. That Count included 2 Yellow-billed Loons, 1312 Steller's Eider, 2 Parakeet Auklets, and 48 Emperor Geese.

Keep on thinking cold and head for Juneau, peninsular capital of Alaska. Nearby, along the Chilkat River, will be one of the world's more remote wildlife spectacles: the convening of as many as 3500 Bald Eagles. Along the Chilkat River north of Juneau, thousands gather to reap a Chilkat salmon harvest. It seems that warm water wells up into the river allowing the salmon to spawn through late fall and early winter. It's the eagles' thanksgiving dinner and they come from as far away as Washington state to partake from *OCTOBER* through *JANUARY*. View these magnificent birds perching in tree after tree when they are not feasting on the spawning salmon. Fortunately, Alaska has created the Chilkat Bald Eagle Preserve to protect the environment and ultimately to provide wildlife enthusiasts with a rare lifetime experience. What a Christmas present to give your favorite birder!

Ann van der Geld, who reports for the *Newsletter* of the Hawk Migration Association of North America, sought a slightly lesser spectacle at Harrison Bay in British Columbia. She reported that on *JANUARY* 3, 1986, she saw 600 of those magnificent birds. She was so excited that she returned a couple of weeks later and could hardly believe counting 1000, some rising on thermals in kettles.

The northwest Pacific coast is another first rate locale for watching birds in moderate comfort. Forested areas are closer to the coastal areas than on the Atlantic side. Varied habitat provides the Pacific birder with a wider variety of species. Loons, grebes, ducks, geese, gulls, and guillemots join the wrens, robins, finches, and sparrows. Pacific coastal areas will be good for loons but your best bet to see the rare Yellow-billed Loon will be along Washington coastal areas, or maybe at Victoria, British Columbia. It will be a red letter day if you spot one. Common Murres will be fairly common in Pacific coastal waters. The San Juan Islands are good for "catching" murrelets and auklets. CBCers in Seattle in 1986 found 121 species during the Count with enormous numbers of some. Western Grebes were well represented with 1329 individuals counted along with four other species of grebe, and there were more Coots than you want to see. Counters spotted 1 Saw-whet Owl, 20 Anna's Hummingbirds, 948 Bushtits, and 4992 European Starlings.

FEBRUARY is the peak time north of Seattle at the Skagit River delta. From early *NOVEMBER* through late winter, flocks of Snow Geese and both Swans, Tundra and Trumpeter (the largest of all waterfowl), gather in another of nature's winter sensationals. Watch also for Gyrfalcon, the largest falcon in North America, best seen from *NOVEMBER* to mid-*FEBRUARY*. East of nearby Mt. Vernon, visit the Skagit River Bald Eagle Natural Area in mid-*JANUARY* to view what may be the largest aggregation of eagles in the lower 48 states.

FEBRUARY is a fine time to follow in the path of Lewis and Clark along the Columbia River in Oregon. The "L & K" National Wildlife Refuge is the largest marsh in western Oregon and thousands of Tundra Swans and Canada Geese gather there in *FEBRUARY* and into early *MARCH* preparatory to their northward migration. All along the river, from its mouth up to Portland, offers the enterprising birder lots of avian activity with up to a couple of hundred thousand waterfowl spending the winter in this mild climate.

RUN FOR THE RARITIES

Like every other season, winter produces exciting birds that tickle the fancy of both veteran and neophyte birder. The veteran may just make an extra effort to see a rarity; for the neophyte, it may be serendipity.

Real rarities can show up when you least expect them. World birder Phoebe Snetsinger's winter wonder (a real Christmas present) was practically in her own backyard. It ended with the Horned Guan, Harpy Eagle, and Zigzag Heron on the list of her top 10 world birds. (She's obviously in the veteran class.)

> The whole episode of the finding and eventual identification of the mystery gull in St. Louis and its subsequent cooperative behavior was a dramatic scenario which we can never hope to repeat. I love gulls and identification challenges; the icing on the cake was that when we solved the mystery, it turned out to be an extraordinary rarity—a happy conclusion for many and certainly the local bird of a lifetime.

That Slaty-backed Gull was way off-course; it belongs along the Pacific coast of northeast Asia and coastal China! It was just the second North American record south of Alaska.

A Rustic Bunting (*Emberiza rustica*), relative of North American buntings and sparrows, showed up in Kent, Washington, in *DECEMBER* 1986. It came from northern Old World forests. Watch all those little sparrowlike birds, and check carefully for one that looks just a little different from the rest. It was the Eurasian Siskin that attracted so much attention in *FEBRUARY* 1988. The North American Rare Bird Alert *Newletter* reported that more than 400 alertees showed up in Toronto to see it. That audience, however, pales in the light of the 750 lusty listers who ventured to Ventura, California, to check off the Xantus' Hummingbird on its first visit from Baja to the states.

A bit easier to see was the first Crane Hawk seen in North America (north of the border). Texas caught the brass ring on that one in *DECEMBER* 1987. It was a brass ring too for Steve Perry of Campbell, California for it was the species that landed him one ahead of Benton Basham's Big Year total of 710. That hawk, distinguished by its long orange legs, was both a stunning and cooperative bird. Its normal range is down Mexico way and farther south to Argentina.

Texas, the big *T,* rates a big *R* for rarities of record. The winter of 1988 was pretty special. In addition to the Crane Hawk, the North American Rare Bird Alert alerted members to a Gray-crowned Yellowthroat (Mexican cousin of the Common Yellowthroat), Nutting's Flycatcher at Big Bend, Tropical Parula, Masked Booby, and White-collared Swift at Freeport, in addition to the goodies mentioned earlier.

Proving that Texas and the coastal areas don't have a corner on the rarities market, a Falcated Teal (which might be mistaken for a Green-winged Teal) visiting from Asia was seen in *DECEMBER* 1986 near Sandusky, Ohio. Try this from the NARBA *JANUARY* 1987 *Newsletter*:

> To usher in the new year, a white-phased Gyrfalcon was seen swooping down to help himself to one of several Ivory Gulls that were feasting on a seal carcass in northern Newfoundland.

These two species would be very special on anyone's life list.

Winter does usher in the new year, both the calendar year and the birding year. By *FEBRUARY,* the end of the birding year, spring is beckoning. In Florida's Big Cypress Swamp, the cypress trees are greening. Shorebirds are beginning to respond to some inner anticipation of their new year. The resting season is ending, the nesting season awaits.

Seven

A Place for All Seasons

*B*irding Mississippi during the hot, still summer is not likely to provide much birding excitement. Alaska in the dead of winter is better for dog sledding than dogging birds. Areas along major flyways can be very exciting in spring and fall. Where you are when, does make a difference. Weather also makes a difference; birding can be exciting or dull during migration, depending on how the weather is. In fine sunny weather, migrating flocks may customarily overfly very large areas, only dropping down in the face of buffeting storms. Summer concentrations of nesting birds in many places across the continent can provide exceptional birding experiences. Summer and winter may be only so-so where you are.

Some places boast good general birding year-round. Some may be legitimate "hot spots" attracting birders in great spurts. Others provide satisfying experiences season in and season out. They may have resident species that get a "good bird" medal at any time of year rather than getting rave notices for some particular time of year. Don't be put off by visions of clouds of mosquitoes or black flies, junglelike humidity, bitter cold, or biting winds. Some of this just goes with the birder's territory. Experienced birders learn to grin and bear it if they can just get a good look wherever, whenever, at the Bristle-thighed Curlew, LaSagra's Flycatcher, or another most-wanted bird.

Some generalizations can be made to provide a rough guide on areas that will reward the birder at almost anytime of year. Coastal areas generally will be good, particularly if winds and currents modify climatic extremes such as in the Pacific Northwest, though they tend to be best in spring and fall. States and provinces that have great habitat diversity attract birds adapted to specialized climate and terrain. Mountains, deserts, seashores, lakes, or wetlands each attract particular bird species. Some species like it hot, others cold; some like it high, others low; some like it dry, others wet. If you would see large numbers of species on your trip across the country, plan to take in a variety of habitats.

SUPER-STATES

States and provinces known to have large populations of species are more likely than not to provide good year-round birding. California and Texas beat the pack by a bunch, running neck and neck. The American Birding Association keeps track of these statistics and comes up with the following states and provinces with 400 or more species.

Texas, 579
California, 575
Arizona, 502
Florida, 475
New Mexico, 458
Oregon, 454
British Columbia, 452
Massachusetts, 451
Colorado, 441
Oklahoma, 440
Ontario, 436
New York, 435
New Jersey, 434
Nevada, 431
Alaska, 430
Louisiana, 429
Kansas, 421
Washington, 420
Nova Scotia, 414
Nebraska and North Carolina, 410
Illinois and Minnesota, 406

Not only is this listing of super states two-thirds higher than it was in the previous year, most of the states posted higher totals. It is likely that such increases reflect increased numbers of birders in the field and increased care in identifying uncommon species. Perhaps a few off-shore winds helped too.

The two front-runner states offer some of the best of year-round birding adventures. Arizona is another place where lots of birders choose to live because they can pursue their interest around the year, and around the clock, too. Believe it or not, Florida is not only a winter vacation spot but a top tourist destination nearly year-round, and the state offers special birding treats regardless of season. Massachusetts has long been known for exceptional birding and for expert birders.

Birds, take no note of state, provincial, or even national boundary lines. It's just convenient for us to think in these categories as we do in this book. Comments about species of a given state or province may well be true of neighboring states or provinces. Guidebooks for a particular area may note a specific place for seeing a particular bird without mentioning that the bird may better be seen across the border of the neighboring county, state, or even country. Many of the species for which Arizona is famed may also be seen in New Mexico. So if your tracks stop in New Mexico for some reason, be not disheartened. Mexican species cross the Rio Grande with impunity. Canada and the United States share many species. Then there are those rarities that reach the continent from South America, Asia, Europe, Africa, and elsewhere, some visiting quite regularly. Experienced bird tour leaders have identified those areas that offer the best combination of species regardless of political boundary lines.

CULLING CALIFORNIA

California provides a year-round menu for the voracious birder. Theoretically, if you looked long and hard enough and were able to see all the species of birds known to California, you would see 65 percent of all the species listed for North America. Visiting California for any reason is a good excuse to take your binoculars and a couple of bird guides no matter what the time of year. California, known for its offshore waters, is on the Pacific flyway so it's a good place to be during migration. Remember that flyways aren't narrow straight lines but often cover broad areas up to a couple of hundred miles wide. In summer, some species fly north from Mexico for a postbreeding look around the territory. Many northern nesters move south into the state for the winter. Southern California is felicitous for both birds and birders.

In addition to migratory species, there are resident Mountain Quail, Black Oystercatcher, Band-tailed Pigeon, White-throated Swift, Anna's Hummingbird, Nuttall's Woodpecker, Black Phoebe, and Sage Sparrow, not to mention the two "tits": Bushtit and Wrentit. Neither are related to the familiar Titmice. The delightful, long-tailed Wrentit isn't related to anything else but technically is part of the family Muscicapidae—Old World warblers and flycatchers. The Bushtit boasts better lineage, being a member of the family Aegithalidae which includes the Long-tailed Tit, well-known in Europe and Asia, along with some other bushtits from some fairly exotic places in Asia.

California is also home to other year-round residents for which the state is famous, one of whom is Arnold Small, author of *The Birds of California*. Small identifies 164 species that live in the state year round. "Flagbird" of that list surely is the California Quail. It carries the name of the state, and even has a prominent flaglike plume. Trailing the quail might be the California Thrasher. Although the quail is widespread through western coastal states and, in fact, has invaded a number of places around the globe, the thrasher can be found only in a limited area of California and down a ways into Baja.

Many species that birders particularly want to add to their lists can be found some place in California anytime of the year. Just drive away from the Los Angeles airport and look carefully at the doves around the palm trees: one will doubtless be a Spotted Dove. Band-tailed Pigeon, Burrowing Owl, Costa's Hummingbird, Nuttall's Wood-pecker, Rock Wren, Le Conte's Thrashers, and Hutton's Vireo are others. California is the only place where you will see California's en-demic Yellow-billed Magpie, and then principally in the Sacramento valley. If you want to see this saucy bird, just go where it hangs out, anytime of the year. The Tricolored Blackbird is another near endemic of this area—a few are found in Baja. Although you could see it any-time of the year, like most birds it's easiest to see during breeding season in the Central Valley and along the coastal areas from Sonoma County south. It is not widespread, so seek local advice on the best place to find it.

California's tremendous geographic diversity is what makes it such an exciting place to live. "I can ski in the mountains in the morn-ing and go surfing in the afternoon" is a familiar theme. Birds are attracted to the differing habitats that meet their differing needs: from the high Sierra Nevadas down to the beaches and farther down to be-low sea level at Salton Sea; from redwood forests of northern coastal areas to hot deserts in the southeastern part of the state. Small identi-fies 25 major habitats.

In addition to most any place along the coast, there are 64 na-tional wildlife areas, national parks, monuments, forests, and so on that provide fantastic opportunities for birding. The California Fish and Game wildlife areas, state, county, and city parks, along with pri-vate parks and gardens, truly make the state a year-round paradise for the live-in or traveling birder. For sea birding, there are pelagic trips scheduled every month of the year although most depart from *AUGUST* through *NOVEMBER*. During migration, the gentle cliffs cradling Mon-terey Bay are good lookout points for pelagic birds, and you don't even have to get in a boat. You will delight in the harbor seals, the sea lions, and the playful sea otters floating on their backs, cracking abalones and sea urchins on their tummies. Plan on time to visit the Monterey Bay Aquarium, a stunning research and tourist facility. Reports of whale sightings on your trips with Shearwater Journeys likely will be contributed to the Aquarium. It would be a good place to visit for a preview of what you may see out on the waters.

Time permitting, bird nearby coastal areas. Spotting other birders at Moss Landing just a few miles north of Monterey, you will likely find they will be on the boat trip with you. Wander south to Carmel, that lovely arty village of yore. When you tire of shopping, drive south of town to the beach. One late *AUGUST* day, Elegant Terns were massed on a sandy spit, rosy hued breasts matching the delicate tints of the sunset sky. Follow on down Highway 1 to Henry Miller's Big Sur. Withstand temptation and only stop at every *other* overlook. But do stop, peer down the jagged cliffs, and watch the brown kelp beds undulating on the sea's surface. Scan the rocky beach below and watch for a Wandering Tattler teetering on the wave washed stones, probing around them for a tasty supper.

If whale watching from a bouncing boat is not your cup of seawater, head down to San Diego. Drive out to Point Loma and watch the Gray Whales blow past during their migration. While watching whales, Eastern birders will be watching for western species: Glaucous-winged, Heermann's, and Mew Gulls. Eastern and western visitors alike will be watching for the uncommon Thayer's Gull. Though some might take umbrage, the Lane guide to southern California says, "No other major city in the United States offers such outstanding opportunities for birding as does San Diego." You will be well-rewarded by following the Lane guide. Another guide indispensable to the wayfaring birder is *The California Coastal Resource Guide* (University of California Press, 2120 Berkeley Way, Berkeley, CA 94720). It will take you to every nook and cranny along the state's birdy coast.

A short, generally easy, pelagic trip goes from Ventura out to the Channel Islands National Park. You will see seals on the rocky shores of Anacapa Island and have a chance to clamber over the summit of the small island. The islands are important nesting areas for both northern and southern species, the only nesting area for Brown Pelicans along the west coast of the States. The Xantus' Murrelets (honest, they are real birds) colony on Santa Barbara Island may be the largest in the world. One incipient world birder learned on this introductory trip that a *pelagic* didn't require a prescription.

One of the big advantages of California as a birder's dream world, is the proximity of splendid wildlife areas to San Diego and other major cities. The Santa Monica Mountains and nearby coastal areas north of Los Angeles will provide excellent birding opportunities no matter when you may be visiting. The very active Los Angeles Audubon Society will be glad to have you join one of their field trips. A trip to Antelope Valley is a good one for it has a wide variety of both water and land birds, in addition to nonbird species like coyotes.

Point Reyes National Seashore north of San Francisco not only is a beautiful place to visit, it is rich in land and seabird life anytime of year though it may be foggy in the summer. From there you can spot the Farallon Islands that host many of the Alaskan breeders such as the Tufted Puffin. Call the Golden Gate Audubon Society for information on boat trips out to view the islands' great colonies of seabirds. Thousands of Common Murres nest on the rocky cliffs.

On your way there, meander along Highway 1 and stop as frequently as you can along the shore of Bolinas Bay. Ideally, visit the Audubon Canyon Ranch with its great heronry, active in *APRIL* and *MAY*, and its woods and canyons. Walking through the woods you might spot a Spotted Owl. From an overlook, you see Pacific waves, spot elephant seals lolling on the beach, and notice visible evidence of the infamous San Andreas Fault.

Go south of San Francisco along the coast to the Año Nuevo State Reserve. Watch for Harlequin Ducks and the coveted Black Swift. *JULY* is probably the best time. Black and Surf Scoters, Pigeon Guillemots, plus Brandt's and Pelagic Cormorants are best seen in the spring. It's a famous seal breeding beach and if you know your pinnipeds, you could see as many as four species. Not far south of San Francisco, but on the inland side of the coastal mountains is a lesser known but valuable marshy wildlife area, the Palo Alto Baylands Reserve, utilized by a vast majority of migrating water birds for resting and general rejuvenation.

The rich Central Valley area is host to both migrant and resident birds. A string of wildlife refuges are flooded in the fall and winter to provide resting areas for migrating water birds, Long-billed Curlews, and Sandhill Cranes. At the southern end of the state is the salty Salton Sea, one of the most famous birding areas in North America. Its rank with Death Valley as the hottest place in the United States is well deserved. It is *hot* in the summer, but the stalwart will be there to see the rare Yellow-footed Gull, and some seabirds such as Parasitic Jaegers not generally found over land.

North of the Salton Sea is the Big Morongo Wildlife Reserve, a desert oasis worth a visit at most any time of year. Migrating birds moving up from western Mexico pass through this area in *MAY* and return in the fall, while many stay to nest. It's one of the few nesting areas for Brown-crested Flycatchers in California.

Desert species flourish in the Mojave Desert as they do in nearby Nevada and Arizona. No North American list would be complete without checking off a Chukar. *Alectoris chukar* was introduced from Asia as a gamebird but, as is the case with several other members of the partridge family, has settled down in North America and is an official "tick." Other good desert species include Bendire's and Crissal Thrashers, Black-throated and Black-chinned Sparrows, and Black-throated Gray Warbler.

TARRY IN TEXAS

Texas is a good place to bird at your own pace, in your own manner. Its abundant year-round bird life is legendary and it can be spectacular in the spring with migrating birds moving through, summer residents arriving to nest, and resident birds singing their little hearts out. A rainbow of wild flowers along the exceptionally well-maintained roadways and an exceptionally wide range of habitats help make Texas a birder's paradise. There is a Peterson guide to Texas birds as well as two Lane guides. That's a sure sign of avian eminence.

One of the best Texas areas is in an arc along the Gulf Coast from the far western Louisiana coastal area through Beaumont, down through the Houston area to Brownsville at the southernmost point of the state. Although the avid birder will find guides for some of the specific spots, the Lane guide, *A Birder's Guide to the Texas Coast,* is used almost universally by knowledgeable birders. This guide lists 11 principal points for birding with uncounted references to specific spots along the way. A second Lane guide, *A Birder's Guide to the Rio Grande Valley of Texas,* covers another arc from Brownsville, up across the Edwards Plateau, down to Big Bend National Park, ending up in El Paso.

Every season in the largest of the lower 48 states offers tantalizing birding. Straddling the 100th meridian, eastern and western species converge here making it a particularly challenging area. This is where you begin to see more Swainson's and Ferruginous Hawks than Red-shouldered Hawks. This is about as far east as the Prairie Falcon shows up. You may see both Eastern and Western Screech-Owls, Eastern and Western Meadowlarks, plus Eastern and Western Wood-Pewees. This general area is where the westbound traveler loses the Ruby-throated Hummingbird and has the potential of seeing others, where Red-bellied Woodpeckers stop and Ladderbacks begin, where you begin to see both races of the Yellow-rumped Warbler, Myrtle and Audubon's, and where you really have to look carefully to differentiate the Boat-tailed and the Great-tailed Grackle.

Waterfowl and other birds coursing the Central flyway pass through the heart of Texas. For birds flying north from Latin America in the spring, the arc of its Gulf Coast is the first welcoming land sighted on their long journeys, just as it is the end of the continental journey for those species flying south in the fall to find comfortable winter accommodations.

The southern tip of Texas is almost as far south as the only subtropical area of the continent north of the Mexican border: south Florida. Winter weather is sunny and warm in both places. The Rio Grande River is but a narrow strip separating Mexican species from those

Greater Roadrunner "beeps" across western landscapes

north of the border. Ringed and the little Green Kingfishers and other primarily Mexican species regularly cross the border year-round.

Birders who have never visited Texas will come away with an impressive list of birds no matter what time of year they choose to visit. White-faced Ibis will be found along the Gulf Coast. Black-shouldered Kites, Harris' Hawk, and Crested Caracara are good bets. Plain Chacalacas hold forth at Bentsen State Park and Scaled and Gambel's Quail and Wild Turkey should always be found some place in the state. Inca Doves and Greater Roadrunners like the same dry habitat and, along the lower Rio Grande, try flushing a Common Pauraque. While there, watch for Green Jay, Great Kiskadee, Altamira Oriole, Olive Sparrow, and Buff-breasted Hummingbirds, the little one possible all year but best seen from spring until early fall. Verdin, Bushtit, Cactus, Canyon, and Rock Wrens, plus Brown Towhee and many others will be found some place year-round.

Texas is so famous for specialties that a quick rundown by seasons may be helpful. Table 7–1 shows the species listed by the North American Rare Bird Alert in its 1987 *Newsletters.*

From the woodlands in the east to the grasslands of the west, from the deserts to the mountains, from coastal waters to the muddy Rio Grande, Texas seems to have it all.

Table 7–1 Bird species found in Texas, listed by season.

Species	Spring	Summer	Fall	Winter
Masked Booby			X	X
Muscovy	X		X	X
Hook-billed Kite	X		X	X
Montezuma Quail	X	X	X	X
Yellow Rail	X		X	X
Black Rail	X			
Northern Jacana				X
Ruff		X		
Lesser Black-backed Gull	X		X	X
Little Gull	X			X
Red-billed Pigeon	X	X	X	
Red-crowned Parrot			X	X
Green Parakeet			X	X
Ruddy Ground Dove			X	
Ferruginous Pygmy-Owl	X		X	X
Mexican Crow				X
Rufous-backed Robin	X			
Clay-colored Robin	X		X	X
Tropical Parula		X	X	
Gray-crowned Yellowthroat				X
Golden-crowned Warbler				X
Crimson-collared Grosbeak			X	X
Blue Bunting				X
White-collared Seedeater	X		X	X

An X does not indicate that the species is found throughout the state, during the season indicated; or that it can be found for the duration of that season; or even that it's not found at other times of the year. It indicates only that the species was found during at least one month in the season indicated.

AROUND THE YEAR IN ARIZONA

Arizona is a deservedly famous state for year-round living and visiting. Traveling families head for the Grand Canyon for summer vacations, while snowbirds from the snowy north seek out the sun around Phoenix and Tucson in the winter. American Indian culture thrives there and national forests, wildlife refuges, mountains, and deserts offer a paradise to birders. Even if you can't visit the state right away, you can read *Arizona Highways* magazine and derive vicarious pleasures from the spectrum of intriguing perspectives on the state.

Although southeast Arizona is world famous for its spring and summer birding, that area delivers superb birding year-round. Fall migrants begin flying south in late *AUGUST* and on into *SEPTEMBER.* Sparrows that are absent in the summer, are seen not only during spring and fall, but they stay there in winter. The Sage Thrasher may occasionally be found in winter at lower elevations. Hutton's Vireo likes it there year-round. No wonder many topnotch birders choose to live there.

Southeast Arizona, that area of the state south of Tucson to the Mexican border and east to the New Mexico border, includes the Nature Conservancy's Mile Hi Sanctuary and a rich trove of other good birding areas, particularly in the Chiricahua Mountains. The Mile Hi reaps deserved acclaim for many birding jewels besides the astounding variety of hummingbirds for which it is most famous. Listen to this list of oft-wanted birds found there year-round: Scaled, Gambel's, and Montezuma Quail; Wild Turkey; Band-tailed and White-winged Pigeons; Inca Dove; Barn, Western Screech, Whiskered, Great-horned, Spotted, Northern Pygmy, and Burrowing Owls; White-throated Swift; Gray and Vermillion Flycatchers; Strickland's Woodpecker, Mexican Jay, Chihuahuan Raven, Bridled Titmouse, and Common Bushtit; Curve-billed and Crissal Thrashers; Black-tailed Gnatcatcher, Phainopepla, Olive Warbler, Painted Redstart, Pyrrhuloxia, Lesser Goldfinch, Green-tailed, Brown, and Abert's Towhees; Rufous-crowned, Cassin's, Black-throated, and Rufous-winged Sparrows; and Yellow-eyed Junco. That's some list!

If it seems to you that owls predominate on lists of Arizona, you are right. According to Kenn Kaufman and Kate Stenberg, writing in the March/April 1988 issue of *Birder's World,* there are 13 species of owls to be found in southeast Arizona including the sparrow-size Elf Owl, the smallest of the world's owls. It likes insects, so is found there in late spring and summer.

Furthermore, southeast Arizona is the only place north of the border where some of these species can be found. Those wondrous birds won't just parade by you. Hiking into the mountains and down to the valley floor, along with patience and time, should reward you with most of them.

Anytime of year you are in southeast Arizona, a visit to the Arizona Sonora Desert Museum is an absolute must. In fact, do it as early as possible during your visit. Now any birder worth his or her binoculars knows that the only birds likely to be found in museums are dead ones, laid out in neat rows in wide drawers. Not so the Desert Museum, a living, outdoor, interpretive exhibit of the bird, mammal, and plant life of the huge Sonoran Desert that covers parts of Arizona and Mexico. Not only is it a tourist attraction of distinction, it is an attraction to the wild birds of the area that flock around to help the caged animals clean up the breakfast crumbs. The interior race of the Brown Towhee with its little necklace of spots is a frequent visitor. Such wild birds can be ticked off on your checklist.

SUNNING, TANNING, AND BIRDING IN SOUTH FLORIDA

You might not think of south Florida as the best place to fog up your binoculars in the summer. The weather is more pleasant during migration, or during the winter when, for Canadians and those from the northern states, there are the multiple objectives of seeing birds, acquiring tans, and avoiding snow—in whatever order fits your priorities. Like many areas in coastal Texas, south Florida is great for the big birds—herons, egrets, ibis, Wood Storks, and Roseate Spoonbills. Most visible in winter and spring some will be seen throughout the year. The Florida race of the Sandhill Cranes are resident birds and each winter welcome the migratory Greater Sandhills.

Special species attract attention in the summer making this part of the world good for year-round birding. The Gray Kingbird is a good example. This summer breeder, a visitor from the West Indies and northern South America, is easily seen sitting on wires in the greater Miami area and over on the west coast in places like Sanibel Island. The Black-whiskered Vireo is more shy and retiring as is the Mangrove Cuckoo. Swallow-tailed Kites are fairly common summer soarers. From *APRIL* through *OCTOBER,* the Antillean Nighthawk, a new species split a few years ago from the Common Nighthawk, can usually be found down on the Keys at airports in Marathon and Key West. Be there early in the evening, and learn its call as dusk makes identification from field marks virtually impossible.

Sanibel Island, particularly the Ding Darling Wildlife Sanctuary, is a good place to bird year-round. There are migratory birds in both spring and fall. The ponds in the Sanctuary are so close to the road (and vice versa), that it's a great place to photograph as well as scope the Roseate Spoonbills. You nearly always see at least one Reddish Egret. Watch their bizarre splashing in circles, wings spread, long beak jabbing the water's surface. It's a behavior known as *canopy feeding*. The Reddish thinks the little fish swim to hide in the shadow its wings make. Maybe they do. Least Terns and Black Skimmers are summer nesters in Sanibel and in other coastal areas. Hot? Yes, but along both coastal areas in south Florida, a lovely breeze is generally blowing. At least that's what those of us who live here tell the visitors on hot, muggy, breezeless days. Bugs? Yes to that too. The "no-see-ums" can be annoying at unpredictable times. But then many other places have their special nuisances.

Many south Florida specialties can be found year-round. Birds like the Red-whiskered Bulbul are found only in a small area south of Miami, but the Spot-breasted Oriole, though scarce in the dead of winter, has a more widespread range. The White-crowned Pigeon can be found from south Florida through the Florida Keys. Almost anytime of year is a good time for Limpkin, Chuck-will's Widow, Smooth-billed Anis, and Purple Gallinules, those iridescent lily-padders that attract superlatives from the tourists. Birders seeking these species for their North American list will get down to the huge peninsula whenever they can. Winter may offer more comfortable weather, but summer offers more comfortable rates.

WORLD BIRDING AT HOME

South Florida is also the place where you can become a world birder without leaving home—literally for those of us who live here. The greater Miami area is the exotic bird capital of North America at anytime of the year. The Tropical Audubon Society sponsors a limited number of trips to shopping center parking lots, backyard feeders, and Casuarina tree roosts to see the local avian exotica. The extent of international bird life is indicated by such a one-day trip in the Miami area in November 1988 as reported by Bruce Neville, official lister:

> Green Parakeet, Mexico and Central America
> Chattering Lory, Moluccas in Indonesia
> Yellow-headed Parrot, Mexico
> Yellow-crowned Parrot, Panama
> Jungle Mynas, Southeast Asia
> White-fronted Parrot, Mexico and Central America
> Red-lored Parrot, Central America and northern South America

None of these species is on the American Birding Association list. The Canary-winged Parakeet from South America *is* on the list along with the Red-whiskered Bulbul, originally from Southeast Asia. Neville estimates that 45 to 50 members of the parrot family have been sighted from time to time in the greater Miami area, and that the total of all exotics is around 75 or so. Mort Cooper, who frequently escorts visiting birders around the area to see the specialties, says that most visitors are interested in seeing the "countable" species, but that they are fascinated by the great variety of exotics. "Put them on your 'escrow' list," is Mort's advice.

World birder visits Homestead, Florida, to spot Collared Doves

A number of today's exotic species doubtless will become established as breeding populations in future years and may be added to the North American list. Phoebe Snetsinger, who treks about the world looking at birds, was seeking such a gratuitous "tick." Stopping in Miami on her way to Brazil, she visited the Homestead City Hall, not for the purpose of paying taxes or obtaining a zoning permit, but to see the large flock of Collared Doves to which the city fathers were giving free rent. At the time she ticked off the species, it wasn't yet officially on the North American list but she thought it might soon be added.

Some local watchers believe the Monk Parakeet should be on the official list since hundreds of Monks are now exhibiting their nesting and breeding proclivities by building and rebuilding condominium-type nests in palm trees at numerous locations throughout south Florida. Red-crowned Parrots from Mexico are often seen in the greater Miami area but serious birders don't think it sportsmanlike to list them unless they are seen in south Texas. For similar reasons, the Budgerigars, the familiar "Budgies" native to Australia but occasionally seen in the Miami area, are only "official" when seen in the Tampa Bay area. This listing business does get complicated.

My own "yard list" of exotics includes Hill Mynas from India, Yellow-headed Parrots from Mexico, Monk Parakeets from Argentina,

Fischers' Lovebirds from Africa, Blue-crowned Conures, Blue-and-yellow and Chestnut-fronted Macaws, and Canary-winged Parakeets from South America, a Cockateil from Australia, and Orange-winged Amazons from Amazonia. For six months, a Salmon-crested, or Moluccan, Cockatoo from Indonesia dined at the seed tray.

The Key West Quail-Dove from Bahamian and Caribbean islands is listable. Birders from all over North America converged on the Everglades National Park in 1979 and again near Marathon in the Florida Keys in 1987 to observe the rare North American record sightings of this beautiful bird of the forest floor. Finding it is a challenge because it is not easy to spot as it walks noiselessly through the shadows. Don't forget, too, that the Cattle Egret, now common in so much of North America, first hit the new world in Florida in the early 1940s. So the Red-fronted Conure (northern Venezuela to Peru) sighted in the palm trees on Key Biscayne may some day be a "good" bird.

The Florida peninsula also catches a good share of strays or accidentals at anytime of the year. Fly-ins from the Bahama Islands include the Bahama Woodstars, Swallows, and Mockingbirds, all seen with some regularity, as are LaSagra's Flycatchers which also belong mostly in the Bahamas.

EAST COASTING

Coastal areas generally offer good year-round birding as the proximity to the oceans not only moderates the climate, but the ocean winds sometimes bring in exciting species from afar. In addition, North American coastal areas constitute major migration routes. Both coasts host millions of birds every year as they pass through, nest, or rest. Weather may be of more concern on the East Coast, particularly winter weather, but unless the wind is blowing so hard you can't stand up or it's raining so hard your binoculars need windowwipers, try out some of the coastal areas up and down the East Coast even if it isn't at the time of year for birding to be in the big league class.

Parker River National Wildlife Refuge north of Boston, with 301 species seen there in the past ten years, attracts attention year-round although it is best known for the spring and fall passage of migrating birds. For birders tied to the North American continent, the general area of the refuge may be the only place and time they will ever see the Ruff, though for the most part it will be the Reeve, the Ruff's "better half," that is commonly seen. Life is truly "ruff" for the Reeve, as it is for so many nondescript females of a species known for its male's spectacular plumage.

Waterfowl concentrations are good in the winter and there's always a smattering of wintering songbirds. There may be an errant Black-headed or Little Gull summer or winter. Black-legged Kittiwakes are occasionally seen anytime of year. Monomoy National Wildlife Refuge, just off the elbow of Cape Cod and accessible only by boat, provides similar year-round birding. In fact, most of the well-known wildlife areas in Massachusetts, particularly along the coast, are good for year-round birding and year-round birders. Along the side of the road overlooking some marsh or mudpond, a line of scopes with intent birders attached is a signal that something interesting has been spotted.

Brigantine, north of Atlantic City (be sure to refer to it as "Brig"), is a wildlife refuge heavily visited by birders from the metropolitan New York area as well as from around the world. As many as 35 species of shorebirds settle down on marshes and ponds in *MAY* and *AUGUST*, easily scopable from a loop road around the diked ponds. Deciduous woodlands and marshes provide a year-round tapestry of birdlife.

From Delaware down to Georgia, wildlife reserves cater to wintering waterfowl. Winter storms, of course, may limit birding opportunities. The wind blowing across Lake Mattamuskeet in North Carolina can be nose-numbing but the National Wildlife Refuge is spectacular with thousands of Canada Geese, Tundra Swans, and more than 22 species of ducks. Just try to be there on a nice day.

Chincoteague in Virginia, famous for both birds and ponies; Presquile, an island refuge at the mouth of the James River, accessible by ferry in decent weather; and the Outer Banks in North Carolina are favored birding areas. Drive down this string of islands hanging like a necklace off the North Carolina coast and spend some time at the Pea Island National Wildlife Refuge. There's something going on year-round. Some wildlife refuges do an excellent job of alerting visitors to what's going on at what time of the year. Here's a month-by-month run-down from the Pea Island reserve:

JANUARY	High concentrations of ducks and geese. Northern Harriers and kestrels fairly common. Herons, egrets, ibis, and shorebirds easily seen. Barn Owls search the marsh for food at dusk.
FEBRUARY	Waterfowl populations high.
MARCH	Spring shorebird migration increases. Osprey begin nesting activity.
APRIL	Shorebird migration continues. Wading birds begin to establish rookery sites.
MAY	Terns return and begin courtship and nesting activities. Willets nesting, osprey hatching, and duck broods appear.

JUNE	Duck broods abundant. Black-necked Stilts protecting nests. Least Terns, American Oystercatchers, Black Skimmers, and shorebirds nesting.
JULY	Ospreys fledge.
AUGUST	Brown Pelican young learning to fish.
SEPTEMBER	Migrating warblers, sparrows, and raptors coming through.
OCTOBER	Migration continues.
NOVEMBER	Gull and Swan populations peak. Black-bellied Plovers and Willets on beach. Peregrines and other raptors often seen.
DECEMBER	Pelagic birds being seen off beach. Barn owls active at dusk. Waterfowl numbers growing. Tundra Swans, Canada and Snow Geese, and some 25 species of ducks settle in for the winter.

That's a good year's birding activity.

Sea trips off the North Carolina coast are a good bet in *AUGUST* and *OCTOBER* when the seas are relatively calm. Look for trips that go out far enough to see what you want to see. Band-rumped Storm-Petrels, for example, are only found far offshore.

Low country coastal prairies of South Carolina, with their waving golden grasses, hang out the welcome mat for wintering populations

Paint lines on pine trees are clues to Red-cockaded Woodpeckers

of raptors and waterfowl. They serve not only the waterfowl winging down the Atlantic flyway but also ducks and geese headed down the Mississippi flyway that veer east to the Carolinas. Cape Romaine north of Charleston, South Carolina, has Tundra Swans, Canvasback, Black Scoter, and Bufflehead in winter, peaking in early *December*; nesting Wood Storks, White Ibis, Wood Ducks, and shorebirds in summer along with 2000 Brown Pelicans; and migrants spring and fall. Except in January when it's too rough, take the boat out to Bull Island for the best birding experience.

The nearby Francis Marion State Forest is a good place to look for resident Red-cockaded Woodpeckers. Watch for old stands of pine trees and look for the two paint lines a couple of feet off the ground. That signifies a woodpecker tree. Santee State Park, adjacent to the Santee National Wildlife Refuge, has good birding habitat including the pine trees infected with heartwood disease favored by Red-cockaded Woodpeckers. One of the largest colonies of these endangered birds can be found in the Carolina Sandhills of northeast South Carolina. If visiting Atlanta, drive down to the Piedmont National Wildlife Refuge, another Red-cockaded area with good year-round birding.

NORTHWESTERN LARKING

Of year-round birding interest, the farthest point from Florida in the lower 48 probably is the Seattle, Washington/Vancouver, British Columbia, area. Any self-respecting birder in that area at anytime of year will head for the Saanich Peninsula just north of Victoria on Vancouver Island, to see the star of the area, the Eurasian Skylark, that "blyth spirit" of poet Shelley's. It's the only place in North America to see this sparrowlike bird with the larklike song. The northwest Pacific Coast is first rate locale for watching birds in moderate comfort even in winter. Coastal areas will be good for loons but your best bet to see the rare Yellow-billed Loon will be along Washington coastal areas or maybe even at Victoria, British Columbia, in winter. It will be a red-letter day if you spot one. Common Murres will be fairly common in Pacific coastal waters. The San Juan Islands, part of that watery northwest corner of Washington, are good for "catching" murrelets and auklets.

Coastal areas and major riverways are honeycombed with a network of wildlife refuges, hostels for migrating waterfowl. Some birds like it so well they spend the winters here. Migrating landbirds use the coastal route and some swallows and western warblers spend their summers nesting and raising the kids. Coastal areas from Vancouver south to California are places to seek nesting Leach's Storm-Petrels, Tufted

Puffins, Pigeon Guillemots, Rhinoceros Auklets, and Common Murres along coastal cliffs or, particularly, on offshore islands. Many of those islands have restricted access to protect nesting sites. Seek expert, local advice on which islands can be approached by boat. Eastern birders will be rewarded along western coasts by Glaucous-winged Gulls year-round. Glaucous means "sea-green," but don't count on color as a field mark. Just study your field guide.

Follow in the path of Lewis and Clark along the Columbia River in Oregon. The "L & K" National Wildlife Refuge is the largest marsh in western Oregon and thousands of Tundra Swans and Canada Geese gather there in *FEBRUARY* and into early *MARCH* preparatory to their northward migration. Along the river from its mouth up to Portland offers the enterprising birder lots of avian activity with up to two hundred thousand waterfowl spending the winter in this mild climate.

British Columbia, with its tremendous variety of life zones, offers adventurous birders a cornucopia from the Boreal Owl in the northern boreal forest, to the Sharp-tailed Grouse of the Grasslands, and the high Arctic pelagic birds who spend their winters in the coastal areas. The George C. Reifel Migratory Bird Sanctuary, operated by the British Columbia Waterfowl Society, is a favorite of knowledgeable birders and offers a year-round menu. Migratory species pass through in large numbers and it is both a major wintering and nesting area. Much nesting activity is observed in *APRIL* and *MAY*, followed inevitably by lots of ducklings and goslings. Shorebirds begin arriving in mid-*AUGUST* followed by waterfowl which reach peak populations in early *NOVEMBER*.

Oregon shares with California one of Roger Tory Peterson's ten favorite birding areas, the Klamath Basin-Tule Lake area, good for year-round birding. In the spring and fall you might see some of perhaps a million migrating waterfowl (try *MARCH* and *SEPTEMBER*). Spring and summer are frantic with all kinds of nesting and nestling activity. Winter is the time to savor the sight of hundreds of thousands of waterfowl. Birds of prey are there also, including the largest wintering concentration of Bald Eagles in the lower 48. By mid-*FEBRUARY* savour the choice of 500 to 600 birds on which to feast your eyes. The big birds will be feasting themselves during the day and then gathering at night in chosen roosting areas.

YEAR-ROUND SAMPLER

Colorado, another of the bird super-states, is a super destination anytime of year. Migrants traverse the state, and high mountain species are the magnet for summer birders. The Denver Field Ornithologists run weekend trips year-round throughout the state, canceling winter trips only for blizzards. They find Canada Geese on virtually any trip any time of the year, but the JANUARY 1987–1988 count number was 11,665. That's a lot of honkers. Denver and Boulder Christmas Bird Counts hovered around 100 species that year. Was it because there were so many birds out there to count or so many birders there as counters? The four races of the Dark-eyed Junco winter over as do the Gray-crowned and Black races of the Rosy Finch. White-tailed Ptarmigan is a particularly good Colorado "tick," recommended in wintertime for appropriate snowy-white effect. Colorado is a good place to look for Pinyon Jays year-round.

All up and down the Mississippi flyway, there are state and national refuges good for passing birds in spring and fall, wintering waterfowl, and summer nesters. If you are aiming for specific refuges, check local regulations. Some do close for periods of time in the winter while others offer limited access during breeding or hunting seasons. Birders in the St. Louis area are active so call on them for assistance. Year-round, that area is famous for the Eurasian Tree Sparrow, known locally as the *ETS,* a neat tick on your North American list.

Winter is cold in Kansas, but in between snow storms, the state can be added to the year-round birding list. The Cheyenne Bottoms State Wildlife Management Area near Great Bend, Kansas, is considered to be the most important shorebird area in the Central flyway and one of the most important in the country for several species. The International Shorebird Survey reports that 45 percent of all shorebirds stop there on their northward journey including over 90 percent of White-rumped, Baird's and Stilt Sandpipers, Long-billed Dowitcher, and Wilson's Phalarope. It's another place to watch for the Sandhill Cranes during migration (it's not far from the main spring sandhill show along the Platte River in Nebraska). In this state, better known for its corn and wheat than its bird life, it is Cheyenne Bottoms with 320 species having been sighted there and nearby Quivira National Wildlife Refuge that helps put Kansas among the birdy super-states. These wildlife areas are best during spring and fall migration though many eastern and western songbirds nest there. All winter, watch at Quivira for Bald Eagles, Rough-legged Hawks, and Prairie Falcon.

Driving through Tennessee? Try the Reelfoot and Lake Isom National Wildlife Refuges, an important stopover and wintering area along the Mississippi flyway. If you are lucky, you won't see more than a quarter million ducks, but winter is a good time for Bald Eagles. Marshes and swamp forests are good summer areas for the wading birds that frequent such habitats. Wander up and down the Tennessee River valley and check out the TVA and other refuges for nature's creatures.

Great Lakes states and the province of Ontario provide plenty of good birding and, although they can in winter discourage all but the most optimistic and hardy birders, they can provide year-round birding worthy of their super-state status. Any time of year while visiting Minneapolis, rent a car and head down the Minnesota River valley. The managers of the MRV National Wildlife Refuge would like to entice you with Tundra Swans in early spring, Northern Orioles and Prothonotary Warblers in summer, waterfowl migrations in the fall, and Bald Eagles in winter. Just don't make the mistake of expecting to see ducks on ponds in late fall. It's amazing how a covering of ice will dissuade them.

Wisconsin, surrounded on two sides by those huge lakes, has some fine wildlife areas used by both migrating and nesting birds. A year-round favorite is the Horicon National Wildlife Refuge northeast of Madison. Although it's especially good in the spring, both spring and fall migration can be excellent. In fact, Roger Tory Peterson called it one of his "top ten" birding destinations. Horicon encompasses the largest freshwater cattail marsh in the United States and lists 248 bird species of which half nest there including two to three thousand Redheads. Lots of other ducks raise their families in the friendly marshes and during spring and fall migration they share the space with spectacular numbers of Canada Geese. Winter is cold but ski and snowshoe trails make the area accessible and Rough-legged and Red-tailed Hawks can be seen overhead, and maybe deer and fox.

Ontario, the Canadian province half again as large as Texas, borders the upper tier of the Great Lakes. Beyond its world famous spring migration "hot spot," Point Pelee, the province has many other well-known areas good for migration, summer nesters, late fall/early winter gull movements, and wintering species such as Evening and Pine Grosbeaks, Common Redpoll, and both crossbills. Toronto birder Helena Wilcox finds she's birding pretty much any time of year.

HAILING HAWAII

A special place on any birder's checklist is reserved for Hawaiian birds. As on so many isolated islands around the world, unique species have evolved over the eons. On the Hawaiian islands there are 39 species and the same number of subspecies that are endemic to the archipelago. Not only does the bird lister value such endemic birds, but on these islands, the numbers of native species that have become extinct, or are facing extinction, lend a sense of urgency to the serious birder.

Native birds, found year-round, are joined from late fall to early spring by migrant waterbirds and shorebirds from North America. The

Great Frigatebird, one of the world's five frigatebirds, but not the one familiar in North American coastal waters, can be seen year-round according to Dan Moriarty, Park Ranger at Kilauea Point National Wildlife Refuge. Anytime is a good time for White-tailed Tropicbird and both Brown and Red-footed Boobies. The Red-tailed Tropicbird is more selective; it is here from *APRIL* to *NOVEMBER*. When your neck hurts from looking high in the sky at the frigatebird, look down in the water for the Pacific Green Sea Turtle and the Spinner Dolphin, a bonus for anyone.

Newell's Shearwater, the Hawaiian race of the Manx Shearwater, can be seen from *MAY* to *NOVEMBER*; the Wedge-tailed Shearwater from *MARCH* to *NOVEMBER*. Go in the winter to catch the Laysan Albatross that you will find from *NOVEMBER* to *JULY*. Moriarty suggests *DECEMBER* to *MARCH* to add the Humpback to your whale list.

Going to Hawaii? Be sure to check the library or one of the book dealers listed in Appendix 2 for a list of those very special Hawaiian species and their location. You won't find these lists on the beach at Waikiki. Topping the list of special species are the honeycreepers, members of the endemic family Drepanididae. So many of Hawaii's special birds are identified as "super-rare" that you might think about going with an organized group. Most of Hawaii's special birds will be found in the Haleakala National Park on Maui, or Hawaii Volcanoes National Park on the island of Hawaii.

Do birds seen on your Hawaiian trip "count?" For anyone, these special birds count for a great deal. For some species, it is probably only a matter of time before they will disappear. For listers who are concerned about what birds are legitimate to count, the answer is both yes and no. If you follow the AOU guidelines, the answer is yes. If you follow ABA guidelines, the answer is, at this writing, no. Those guidelines, however, are under periodic review and it may be that some day the birds of Hawaii will be "legal" counts. H. Douglas Pratt, author of *A Field Guide to the Birds of Hawaii and the Tropical Pacific,* has persuasive arguments favoring inclusion of the Hawaiian Islands. Look back to Chapter 2 for a more complete definition of "what is North America."

North America offers year-round birding for those whose passion is of the avian variety. Birding is good in some places year-round; in many places primarily during the spring and fall migration; in some places primarily in summer. In a few places, some species are better seen in winter than they are at any other time. So choose your season, choose your place, and choose the birds you most want to see.

Appendix 1

A Little Latin and Lingo

*B*irding holds many challenges. It's not enough just to identify the bird in the bush. It's important to know its song, its habits, its Latin name, its family image, and the lingo of the other birders you meet.

PUT THE PIGEONS INTO THE RIGHT HOLES: A BRIEF LATIN LESSON

In 1985, a Shiny Cowbird was sighted in the Florida Keys for the first time in North America. A year later several were seen, and on May 25, 1987, one was identified feeding with some Brown-headed Cowbirds on the grass in front of the marina and visitor's center in Everglades National Park at the southern tip of mainland United States. Distinguished from its companions by an overall purplish cast, a dark eye, and a conical bill, it clearly was a stranger. Although this species was known to the experts, the typical North American birder would look in vain in their North American field guides for a black bird which looked like that one.

Because strays from the Caribbean often invade south Florida, local birders often have a "Bond," the standard guide to the birds of the West Indies authored by the late James Bond. Mine was the 1971 2d edition purchased years ago. Looking up Cowbird, the index listed Brown-headed and Glossy, but not Shiny. Stray birds do come up occasionally from Mexico and Central America. Peterson's Mexican field guide didn't list a Shiny Cowbird either. Ridgely's guide to birds of Panama brought enlightenment. There it was; the plate confirmed the appearance. It's Latin name is *Molothrus bonariensis.* Curiosity led me to look back to the Bond to find out if it might be the same as Glossy Cowbird. The Latin name was the same; it was the same species. In the West Indies, Bond tells us, the bird is also known as Blackbird's Cousin, Corn Bird, Tordo, Merle de Barbade on St. Lucia, and in Martinique, it's Merle de Sainte Lucie. The Latin, or scientific, name remains the same; it is the "language" understood worldwide, no matter what a local name may be. "Latin" doesn't necessarily mean a derivation from the Latin language although some are. However, all such scientific names are Latinized.

World birders know the importance of knowing Latin bird names and terms. For the birder who peers beyond native shores, the international language provided by the scientific names helps enormously in clearing up confusion of different names for different birds in different countries. The typical watcher of birds in North America pays little attention to Latin names of birds, nor to their family lineage. But even here, with the opportunity to spot vagrants from other shores, learning a little Latin and avian genealogy can make birding a more rewarding experience.

Like others of my bird watching friends, I had birded for years before I had learned why birds were arranged as they are in checklists and field guides. We just began to look for gulls in the front part and warblers in the back part. It's often the first question early birders ask.

The arrangement of birds in such checklists and in most field guides is no random plugging of pigeons into pigeonholes. Nor is it "chipped in stone." Scientists have studied birds for centuries. Following the basic classification system established by Swedish botanist Linnaeus in the 18th century, birds have been grouped in phylogenic order. Within the orders of birds, ornithologists have grouped birds according to their physical characteristics. Gulls look and behave like gulls. Those studies continue today and occasionally result in division of species into subspecies, the division of a single species into two or more species, or the combining of previously separate species into a single species based on new knowledge about the birds' special characteristics. (In birding parlance, combining species is known as "lumping"; separating species as "splitting.")

Sound too much like school work? Not for Mary Buck, lifelong Ontario birder. "I have spent many contented hours 'ticking off' my Clements and enjoy the nuances of some of the names, even with my limited Latin. For example tracking down *Buterides striatus,* and figuring out that the reason it was not called Green-backed (or Green) Heron is that they (*striatus* and *virescens*) have been lumped—so one must take one's lumps in this life." Lots of listers "lost" a species when this lumping occurred. A current field guide or checklist will keep you current about such lumps; the splits such as the division of the Western Grebe (*Aechmorphorus occidentalis*) into the Western and the Clark's (*Aechmorphorus clarkii*); and changes in name such as Tricolor Heron (*Egretta tricolor*) from the "old" name Louisiana Heron (*Hydranassa tricolor*).

Peter Alden, author of *Where to Find Birds Around the World* and student of taxonomy, warns of the "frailty" of Latin or Latinized classifications and points out that the experts do change these names from time to time: "Scientific names are living, changing, and traceable, they are not necessarily permanent, and learnable at one time." Take heart!

ORDERING THE BIRD WORLD

Where in the world of birds does the Green-winged Teal fit? And why do hawks and eagles come after teals and ducks in the bird guide, but before sandpipers and plovers? There really is some order in all this, so let's look at the universal classification system used worldwide by scientists to provide a common understanding and language to identify all living things. Within the animal kingdom are mammals, birds, reptiles, fish, and amphibians. All birds belong to a single *class,* Aves. It's feathers that distinguish them from everything else.

Orders are the next step down the zoological ladder. The class Aves is divided into orders, 27 of them according to current thinking. The orders comprise all the birds in the world and are arranged very roughly as scientists think they evolved. Orders begin with Ostrich, of the order Struthiformes, thought to be the most primitive of birds, and end with the order Passeriformes, considered to be the most highly developed birds. Passeriformes constitute an enormous order and contain roughly the same number of species as in all the rest of the orders combined. These "perching birds," small, sparrowlike (Latin, *Passer*) songbirds, are often referred to as Passerines.

Twenty-one of the world's 27 orders are represented in North America. Anyone pursuing birding seriously will quickly see representatives of all the orders, arranged as follows in customary taxonomic, or phylogenic, order. This is the sequence in which you will find them in most field guides. General groupings of species that are known in North America are indicated.

Gaviiformes—Loons (Divers in Britain)
Podicipediformes—Grebes
Procellariiformes—Albatrosses, Shearwaters, Petrels
Pelecaniformes—Pelicans, Boobies, Cormorants, and kin
Ciconiiformes—Herons, Storks, Ibis, and kin
Phoenicopteriformes—Flamingoes
Anseriformes—Ducks, Geese, Swans
Falconiformes—Vultures, Hawks, Eagles, Falcons
Galliformes—Grouse, Quail, Pheasants, Turkeys, and kin
Gruiformes—Cranes, Rails, and kin
Charadriiformes—Sandpipers, Plovers, Gulls, Terns, and kin
Columbiformes—Pigeons, Doves, and kin
Psittaciformes—Parrots
Cuculiformes—Cuckoos, Roadrunners, Anis, and kin
Strigiformes—Owls
Caprimulgiformes—Nighthawks and kin
Apodiformes—Swifts and Hummingbirds
Trogoniformes—Trogons
Coraciiformes—Kingfishers
Piciformes— Woodpeckers and kin

Passeriformes—perching, or sparrowlike birds (half the species of the world)

You probably noticed some monotonous spelling; the names of orders end in *-iformes.*

Family is the next grouping of animals within orders. The names of families end in *-idae.* All woodpeckers in the world (although some are called sapsuckers and flickers) belong to the family Picidae, within the order Piciformes.

There are a great number of families in the order Passeriformes. (The exact number of families is subject to some taxonomic dispute and may soon be changed due to increased scientific knowledge resulting from DNA studies.) Of the 183 families listed in the Clements checklist, 3rd edition, North America (as defined in Chapter 2) has representatives of 64 families occurring on the continent. Most species of these families are native, although some just fly by. Some have been introduced, or have introduced themselves to North America. Sometimes families are divided into subfamilies. Some taxonomists (those people who classify things) divide some of the larger families into subfamilies with names ending in *-inae.* The family Muscicapidae, Old World warblers and flycatchers, is thought by some to be divided into subfamilies: Turdinae, thrushes and robins; Sylviinae, Old World Warblers; and other subfamilies not found in North America. Some authorities, however, treat such subfamilies as full families with sufficient distinguishing anatomical features to distinguish them as "birds of a feather."

Genus comes on the next lower rung of the ladder. Most families of birds contain more than one genus. The genus unifies birds (e.g., warblers within the family according to similarity of characteristics). *Sphyrapicus* woodpeckers look quite different from *Picoides* woodpeckers. American birders who have been at it a while may easily identify a Red-tailed Hawk as a *Buteo* or a Cooper's Hawk as an *Accipiter.* The genuses *Buteo* and *Accipiter* group hawks. Similarly, a Western Sandpiper may be referred to as a Calidrid, a member of the genus *Calidris.*

Species is what birdwatchers watch. It applies to a specific kind of bird (e.g., a Black-tailed Godwit). Vireos are easy. Of the family Vireonidae, all the vireos in North America belong to the genus, *Vireo.* The species Philadelphia Vireo is *Vireo philadelphicus,* the species Solitary Vireo is *Vireo solitarius.* How's that for simplicity? Other members of the vireo family live in South America but they're called greenlets, belong to a different genus, and look significantly different. The species name is always italicized with the genus italicized and capitalized.

Some species are further divided into subspecies depending upon where in the world they are found. Members of the same species evolving in different parts of the world often exhibit special characteristics which enable them to be visually identified as a subspecies or race. If it is important that the subspecies be identified in a field guide, it will have a third latinized word added to its first two names, often giving some clue as to where it is found.

Putting the theory into practice, check out magpies. A common species of the American West is the Black-billed Magpie, member of the family Corvidae and therefore cousin to crows, ravens, and jays. That family is one of many that make up the order Passeriformes. Looking at the Latin name we find it pretty easy; *Pica pica*. Would you believe *pica* is the Latin word for magpie? Well, that tells us something; there must be magpies in Europe. A little sleuthing, and we find that this magpie, *Pica pica,* is widespread in the northern hemisphere. We can't stop there for we've just learned that species in different parts of the world are often of different subspecies, 13 for magpies. The one we've seen in Britain and Europe is the *Pica pica pica,* the one in the American west is *Pica pica hudsonia.* Hudson got his name on a lot of things in North America.

The Yellow-billed Magpie, found only in isolated places in California, is a different species, *Pica nuttalli.* Not all birds called magpies belong to the genus *Pica.* If you're birding around the world, you'll find birds called magpies that belong to different genuses, and the bellmagpies of Australasia belong to an entirely different family.

Even if you never anticipate taking your binoculars beyond North American borders, a world bird checklist is a help in figuring out the family lineage of birds we watch and is an enhancement to the pleasure of learning about birds. The Clements checklist, listed in Appendix 2, is the most popular among North American birders and lends itself to being used as a checklist. For the more studious birders, the Howard and Moore checklist is helpful because it lists all the subspecies.

LATIN SHORTHAND

Some of the Latin shorthand may help fix the names more firmly in mind. Then when some "hot shot" birder suggests that the bird on the wire is undoubtedly a *Myiarchus* flycatcher or when the leader on a pelagic trip yells "Alcid—dead ahead," you'll have some idea of what the bird under consideration looks like.

The following refers to birds found in North America. For your "homework," you can look up a species in your field guide, match it to the Latin term, and use it in a sentence. They are arranged alphabetically rather than the way you would find them in your field guide.

Alcid	Family **Alcidae**, auklets, murres, guillemots
Accipiter	Family **Accipitridae**, genus *Accipiter,* hawks
Buteo	Family **Accipitridae**, genus *Buteo,* hawks
Calidrid	Family **Scolopacidae**, genus *Calidris*
Corvid	Family **Corvidae**, jays, magpies, ravens, crows
Empidonax	Family **Tyrannidae**, tyrant flycatchers, genus *Empidonax*
Icterid	Family **Icteridae**, orioles, blackbirds, meadowlarks, and Bobolink
Myiarchus	Family **Tyrannidae**, tyrant flycatchers, genus *Myiarchus*
Pterodroma	Family **Procellariidae**, shearwaters, fulmars, petrels, genus *Pterodroma,* gadfly petrels
Passerines	Order **Passeriformes**, perching birds
Raptor	Latin for Bird of Prey (usually diurnal, not owls)
Turdus	Genus *Turdus,* thrushes

BIRDER'S LINGO: LEARN THE LANGUAGE

If you didn't know better, you might think that your golfer friend boasting about his "birdie" was a fellow bird watcher. Most hobbies or sports generate a language by which its devotees communicate. Birders are no different. Such a language can often be confusing to beginning birders. Some terminology is colloquial, but as birders meet other birders, the jargon spreads. As you encounter birders on your travels around North America, you may find these terms and phrases helpful. Because British birders often show up at this continent's most favored birding "hot spots," a few translations are provided. More serious definitions will be found in the Glossary. Grateful appreciation is expressed to the many birding companions who have contributed to this list.

COMMON PHRASES

It just flew. By far the most common phrase uttered by birders worldwide.

> *Going away from the boat.* Similar to above.

> *I saw where it flew from.* Easier to see where the bird's been than where it is now.

> *It's at nine o'clock about two feet in from the edge of the branch.* You need to know your "o'clocks" (nondigital) to spot birds. Visualize the tree ahead of you as the face of a clock; train your binoculars on where nine o'clock would be; then move them in towards the trunk about two feet.

Straight ahead, flying right to left. Gives you an idea of where to start looking.

You should have been here yesterday—the trees were alive with birds. Of course today, you're watching European starlings and House Sparrows.

COMMON TERMS

Banding. The attachment of a small identification tag, to a bird's leg so that its travels can be charted as others report the identification number. Birders must have permission from appropriate wildlife agencies to do this. British call it *Ringing.*

Big Day. A race, fairly common among competitive North American birders, where they see how many species of birds they can "tick" in one twenty-four-hour period. It's catching on in other countries. Birders also have *Big Years, Big Sits,* and so on.

Bins. British shorthand for binoculars.

Bird guide. Same as *bird book* or *field guide.* Generally refers to a pocket-sized book which provides brief descriptions of birds, along with drawings or photographs.

Blind. A structure constructed to allow the birder to remain hidden from the bird's view but with peepholes through which to see or photograph the bird. British call them *hides.*

Booming Grounds. An area where grouse display in the springtime as part of the mating ritual. Birds utter a low booming noise as they puff out their breasts. Technically known as a *Lek.*

Broads. Shorthand for Broad-winged Hawk. You can tell you are listening to an experienced birder when such terms are tossed off nonchalantly.

Broken wing display. Not a real fracture, but a subterfuge for leading birders away from a nesting area.

Great Blues. Of course, are Great Blue Herons.

Checklist. List of birds to be found in a particular area with a place for the birder to put a check, or tick, upon seeing the bird. Sometimes checklists are included in a field guide; often they are separate folded lists on card stock which will fit neatly inside the field guide. Extensive checklists often are pamphlet or even book-size. A "check" and a "tick" are almost synonymous; individual listing.

Christmas Count. The annual Christmas Bird Count (CBC) sponsored by the National Audubon Society, in which volunteers all over North America, and occasionally elsewhere, count the number of birds of each species which are seen in the course of a day. Lists are later published and used for year-to-year comparisons.

Chum. Not a good buddy, but bits and pieces thrown overboard on pelagic trips to attract fish which attract birds (e.g., pieces of fish, popcorn, etc., often mixed with smelly oil). Participants suffering from *mal de mer* are also said to be "chumming."

Confusing fall warblers. Roger Tory Peterson coined the phrase; refers to look-alike warblers that have lost their distinguishing breeding plumage. The Peterson North American guides show them on separate pages to help you figure them out.

Dropped. Hopefully does not refer to your binoculars but to the bird that *was* in sight but just dropped down into the bushes.

Fall-out. The dropping of migrating birds onto a land area, generally the first the birds have encountered after a long journey across a body of water.

Field mark. The characteristic marks of a bird seen in the field that help identify it (e.g., black cap, white wingbars). Peterson field guides have pointers to important field marks.

Great Blue. Birder's display their birding sophistication by shortening a species sometimes cumbersome name. This one is a Great Blue Heron. Same for *Little Blue, TV,* and a few others. See also *TV* below.

Flight day. A day when you see a good flight of birds during migration. Particularly applicable to raptors.

Got/have. Among the most useful words in a birder's vocabulary. "What have you got?" "I got three life birds in that one field." Translates as "see."

Hotline. May refer to the North American Rare Bird Alert (NARBA), a national system of finding out, or being notified, of where rare birds are being seen; or to informal systems where one birder simply calls another birder to identify where a special bird may be found. May also be called a "grapevine." See Rare Bird Alert below.

Jizz. A vague combination of a bird's characteristics—size, shape, behavior—that can help the observer identify a bird at a distance when field marks are not visible; particularly useful for seabirds and hawks. May be subconscious identification. Term more commonly used by British birders but becoming widely used.

Kettle. No, not for boiling water. Describes a formation of migrating or soaring birds spiraling around warm thermals.

Kill. Has two meanings: (1) a road *kill* is quite obviously a bird that has had an unsuccessful encounter with an automobile; (2) often used by an envious birder who has looked long for a Kittlitz's Murrelet, as in, "I'd *kill* for one!"

Lane Guide. One of a series of where-to-find guides authored by the late James Lane.

LBJ. Short for "Little Brown Job." Describes a small nondescript bird which probably was not identified. In some circles, known as a *COK,* "Christ Only Knows," or simply a *UFO.* Similar to term *BVD,* "Better View Desired." Quite the opposite is a clear identification of a special bird known as a *BGB,* "Bloody Good Bird," or a *CWV,* "Complete Wipe-out View," for example—both slightly British.

Life bird or *lifer.* Refers to a bird the birder has never before seen (i.e., first time in your life you've seen a Common Pauraque).

Life list. A list of all the birds a birder has seen to date. Inveterate listers will update this list with their "lifers" after each birding trip. To "list" means to keep a list or to check off on a printed list the birds you have seen.

Lumping. Refers to the practice of ornithological arbiters in deciding that what had been two separate species, is now one. Listers "lose" a "tick." *Opposite of Splitting.*

Migratory trap. A place where migratory birds are forced to congregate in abnormal numbers due to surrounding inhospitable terrain.

Mobbing. Might seem to apply to the way birders rush to spot a rarity, but really means the tendency of some birds to scream and fly at predatory birds or even snakes. Crows' or Blue Jays' mobbing noises may signal an owl or a pussy cat nearby.

Need/want. Used to identify a species the birder wants to see in order to list it: "I need a Groove-billed Ani."

Opticon. Anything that appears to be a bird but is actually something else. Haven't you turned a beer can into a Pauraque's eyes in the headlights of your car?

Peeps. General terminology for the small sandpipers heard "peeping," often difficult to identify.

Peterson. As in, "Did you look in your Peterson?" Common practice of using the name of an author as a "thing."

Pick up. Translates to, "Add a species to your list," as in, "I hope to pick up an Ivory Gull."

Pishing. Refers to silly sounds birders sometimes make while standing in front of an absolutely birdless bush. They hope some hiding bird will pop out to investigate the "pshh, pshh, swhh, swhh," thinking it might be an invader. Seems to work better in the spring. Also try kissing the palm or back of your hand. Or try a mechanical "squeaker." May also refer to "pishing in the bushes."

Raft. Refers to a large group of ducks or seabirds resting on the water's surface.

Rare bird alert. In North America, it includes a national computerized, telephone-access system by which birders may find out about rare birds, or be notified of their presence. Many urban areas and states have a tape-recorded message one can call to find out about recent important sightings.

Seven hundred (700) club. Fairly exclusive nonorganization. Refers to those relatively few birders who have seen 700 or more North American species. Ditto *600 club.*

Skin(s). Refers to preserved dead birds, in museum or university collections, studied by ornithologists and serious birders seeking help in identifying species.

Smart bird. Refers to a bird's attractive appearance. Mostly a British term.

Split/splitting. Is the opposite of *lumping,* meaning that a heretofore single species has been divided into two or more separate species. Listers get a bonus bird.

Spuh. Articulation of "sp." short for "species." Used to indicate that the bird was clearly a member of a genus containing several look-alike species such as the *Empidonax* flycatchers but that the species was not determined.

Stickbird. Also *twigbird, leafbird,* and so on. Means the birder has sighted something which looks like a bird but on checking with binoculars finds it has turned out to be just a stick. "Falling leaf birds" are common in deciduous forests, diverting the birder's attention momentarily; occasionally being confused with butterflies, another common distraction.

Stuff. Has two meanings: "There's a lot of stuff in that tree" means, "It looks like there are a lot of birds in that tree." "It dropped into the stuff at the bottom of the tree" refers to weeds, sticks, or rubble on the ground under the tree.

Tick. Mostly a British phrase, only partly translatable into American. To *tick* a bird is generally equivalent to checking the species on a checklist. Using the word as a noun usually means the birder has just seen a life bird: "That was a good tick." In North America, ticks bite.

Trash bird. Humorous referral to a long-sought or rare species that suddenly turns up in unexpected numbers. A first sighting of a Bald Eagle on a birding trip is generally cause for some excitement, but after the tenth sighting, it might be referred to in a kindly way as "the trash bird of the day."

Twitcher. Common British term describing the person who is more interested in "ticking" off a large number of birds than in knowing much about them; twitches with the excitement of seeing a new bird, then rushes off to twitch again.

TV. One of the most commonly used abbreviations in the New World; stands for Turkey Vulture. Acronyms are common among birders. *MD* is the common Mourning Dove. You will hear others.

Vagrant. A bird that has been blown off-course and appears in a locality beyond its normal range.

Wag. What the bird's tail does. Short for any of the widespread species of wagtail. Unless you are birding in Alaska, term won't be very useful. There are lots in Europe.

Wave day. Use the term to describe a whole bunch of birds moving through an area (e.g., "It was a super wave day for warblers.").

Window. Has two meanings: "It's just above the window in the tall tree," means "Look for an opening in the tree branches through which you can see the sky," a *window*. White patches on the wings of soaring birds are sometimes referred to as *windows*. Look in your bird guide at the white patches on the wings of the skuas and some raptors.

Wiped out. Success in seeing all the birds of a category: all three species of Ptarmigan (Rock, Willow, and White-tailed).

Appendix 2

Basic Books and Other
Aids to Better Birding

Where to find birding books, equipment, places to stay, group tours, important addresses? Even experienced birders don't always know.

- I couldn't find any bird books at the book store except the Peterson guide.
- Is there a place to stay at Everglades National Park where I won't be eaten alive by the mosquitos?
- How do I find out about organized birding tours?
- Someone stole my binoculars. Which ones are good these days?

Early birders, and some not so early, often ask questions such as these. Experienced travelers have no difficulty in finding travel books in their local bookstores, but birders may have difficulty in finding a field guide to Hawaii or a checklist of Colorado birds unless they know where to look. Accommodations for traveling birders is generally not a problem in North America but when birding off the tourist track in a place like High Island, Texas, you may want to call for reservations. Travel agents can be helpful, providing all kinds of useful information *except* where to go birding and what tour operators specialize in birding tours. You may be exceptionally lucky and know a travel agent who is an ardent birder, but to date, there don't seem to be many. (They just don't know what they are missing!)

Birders planning birding travels want more information than plane schedules or brochures put out by tourist bureaus. Short of thumbing through the last several years of all the periodicals aimed at the birding public, thumb through this chapter for enough of the basic answers to point you in the right directions.

BIRDING BOOKS AND PERIODICALS

Enthusiastic birders find that leafing through a field guide is like eating one salted peanut. You are hungry for more. Yet, for birders, it's not always easy to find the books you want. Libraries vary widely in the extent of their collections of books about birds. Bookstores notoriously have limited supplies of anything beyond a standard field guide and a cocktail table sized book of Audubon plates, the latter priced in a range that would feed a family of four for a week.

Reference books are essential for any birder beyond the birdbath phase. Fortunately, there are lots of books about birds and more are being published every year. Some "classics" are out of print, but can be obtained from a library, or found in a secondhand book store. Some have become collector's items. Mail order book dealers generally have excellent selections. Some specialize in current listings and some can be very helpful in finding out-of-print books.

Basic books are identified here to help the early birder beginning to explore the North American continent, its nearby islands, and the oceans that surround it. With limited exceptions, the listing does not include the many books covering a single state or smaller region, nor are there listed books about specific bird families. Representative books in several categories have been chosen with preference generally being given to the more recent books when there has been a choice of two similar books.

For quick reference, bird books are identified by major categories familiar to birders: field guides, where-to-find guides, and checklists. Books listed cover large geographic areas or merit attention for particular reasons relating to this work. Useful guides to families and species found worldwide are also included. Some emphasis is provided on books that provide seasonal information.

Although *Birding Around the Year* is devoted to the continent of North America and the selection of books reflects the geography, birders becoming serious about this sport or hobby will want to put their growing knowledge of bird life into the larger context of the world of birds. Accordingly, some basic books relating to the birds of the world are listed under appropriate headings.

When to find birds has been the focus of this book. To locate other books that are organized seasonally, we must recall the series written by the famous American naturalist and essayist, Edwin Way Teale. First published in 1956, the series has been reissued in paperback. Teale describes in a leisurely, delightful manner his search for the seasons, the wildlife, and particularly the birds. The series includes *North With the Spring, Journey Into Summer, Autumn Across America,* and *Wandering Thru Winter.*

FIELD GUIDES

As you can't know the players without a program, you can't identify very many birds without a field guide. A field guide was doubtless the first bird book you purchased when a new bird flew into the feeder, or when you decided you ought to know the names of the new birds you were seeing around the campground. For readers who are testing with this book their potential interest in birding, perhaps a definition would be useful. A field guide is a book to use "in the field" that provides pictoral and written descriptions of the birds. Virtually all field guides include maps showing the distribution of the species.

Experienced birders are selective about field guides. Most like a book of a size that will fit into a jacket pocket or one of those pouches designed for the purpose. It helps if there is a distribution map on the same page with the description, although maps in some guides are located in an appendix. Birders concerned about *when* they will be seeing a particular species will study this distribution map which generally will show breeding (usually summer), winter, and year-round ranges, and sometimes will suggest migration routes. Become familiar with the color-code used in your field guide.

The best plates are those that are drawn to show different plumages, preferably against a background that gives some idea of habitat. Such plates give some idea of characteristic activity or stance. Photographic plates are often beautiful but seldom show varied plumages or capture the "jizz" of a bird as an artist can. Some bird guides include a sonogram, a "picture" of how the song looks, but many birders feel they are hard to translate to an auditory image. Good voice descriptions are imperative.

Pete Dunne, well-known birder, author, and director of natural history information for the New Jersey Audubon Society, wrote "A Guide to the Field Guides" for *Natural History* magazine, (May 1987). He covers not only birding guides but those for mammals and wildflowers. Another good article to review is Kevin J. Zimmer's article "Looking for the Perfect Field Guide" in the Spring 1988 issue of *The Living Bird Quarterly*.

United States and Canada

Farrand, John Jr., ed. *The Audubon Society Master Guide to Birding.* New York: Alfred A. Knopf, 1983. Detailed descriptions with photographs. Divided into three volumes: *Loons to Sandpipers; Gulls to Dippers;* and *Old World Warblers to Sparrows.*

National Geographic Society. *Field Guide to the Birds of North America,* 2d ed. Washington, DC:National Geographic Society, 1987. Slightly oversized, thorough descriptions useful for experienced birders; color plates show characteristic perches and habitats, maps on same page.

Peterson, Roger Tory. *A Field Guide to the Birds. Birds of Eastern and Central North America.* Boston: Houghton Mifflin Co., 1980. This is the bird book that made it to the best-seller list. Color plates with characteristic Peterson field marks, similar species sections, distribution map section, and checklist.

—*A Field Guide to Western Birds.* Originally published in 1961. New edition scheduled for publication in 1990.

Robbins, Chandler S., Bertel Bruun, and Herbert S. Zim. *Birds of North America,* 2d ed. New York: Golden Press, 1983. Widely used field guide with color plates, maps, and sonograms on same page.

Extracurricular

Bond, James. *Birds of the West Indies.* Boston: Houghton Mifflin Co., 1985. (Reissue unrevised.) Well-known field guide covering entire West Indies.

Bradley, Patricia. *Birds of the Cayman Islands.* Published by Patricia Bradley, Box 1326, George Town, Grand Cayman, Cayman Islands, B.W.I., 1985. Beautiful little book, superbly illustrated with photographs by Yves-Jacques Rey-Millet.

Brudenell-Bruce, P.G.C. *The Birds of the Bahamas.* New York: Taplinger Publishing Company, 1975. Small field guide.

Hawaii Audubon Society. *Hawaii's Birds,* 3d ed. (Schollenbergen, Robert J., ed.) P.O. Box 22832, Honolulu, HI 96822, 1984. Inexpensive pamphlet-sized guide lists species and good birding spots, illustrated with photographs.

Peterson, Roger Tory, and Edward L. Chalif. *A Field Guide to Mexican Birds.* Boston: Houghton Mifflin Co., 1973. Well-known and widely used standard guide.

Pratt, H. Douglas, P. L. Bruner, and D. G. Berrett. *A Field Guide to the Birds of Hawaii and the Tropical Pacific.* Princeton: Princeton University Press, 1987. Pratt's beautiful plates distinguish this guide, indispensible for the serious birder heading for the Pacific. Contains an index that can be used as a checklist, and a list of endemics by island. Guide covers most of the major Pacific island groups.

Raffaele, Herbert A. *A Guide to the Birds of Puerto Rico and the Virgin Islands.* San Juan: Fondo Educativo Interamericano, 1983. Paperback field guide with short section on where to bird.

Ridgely, Robert S. *A Guide to the Birds of Panama.* Princeton: Princeton University Press, 1981. An oversized field guide, this book not only covers the birds of Panama but serves as a bridge between the Central American countries to the north and the northern South American countries. Useful similar species sections.

Seas and Shores

Harrison, Peter. *Seabirds: An Identification Guide.* Boston: Houghton Mifflin Co., 1983. An outsized guide with detailed descriptions and plates of seabirds of the world in different phases. Also has maps. Widely used by persons with pelagic passions.

Harrison, Peter. *Seabirds of the World: A Photographic Guide.* London: Christopher Helm, 1987. Handy paperback half of which is devoted to the author's excellent photographs, mostly two views of each bird. Second half contains maps and very much condensed information from 1983 volume.

Hayman, Peter, John Marchant, and Tony Prater. *Shorebirds: An identification guide to the waders of the world.* Boston: Houghton Mifflin Co., 1986. Oversized guide contains excellent plates of shorebirds in several plumages.

Madge, Steve, and Hilary Burn. *Wildfowl: An identification guide to the ducks, geese and swans of the world.* Boston: Houghton Mifflin Co., 1988. Companion to previously listed works. Excellent text and outstanding illustrations by Burn.

WHERE-TO-FIND GUIDES

Where-to-find guides have been published for nearly every "hot" and "not-so-hot spot" in North America. Some such guides provide good seasonal information, others are less useful. Some cover a single locality, others are statewide or regional. Space does not permit listing such local guides but many mail order dealers listed later in this section carry most of them.

Alden, Peter, and John Gooders. *Finding Birds Around the World.* Boston: Houghton Mifflin Co., 1981. This book is out of print but mail order book dealers still list it and can obtain copies for you. It can also be ordered directly from from Peter Alden, P.O. Box 1030, Concord, MA 01742, (U.S. $20 post paid). Very good, occasionally out of date, guide to 111 great birding areas including a dozen-plus in North America.

—*ABA Bird-finding Guide.* (Priscilla Tucker, ed.). American Birding Association, Inc. Available from ABA Sales, P.O. Box 6599, Colorado Spring, CO 80934. Detailed information on where to find birds along with detailed maps contributed by members of ABA. Arranged by state in two loose-leaf binders. Updated inserts provided to members.

Edwards, Ernest Preston. *Finding Birds in Mexico.* Sweet Briar, VA: Ernest Preston Edwards, 1968. Supplements published periodically. The original guide along with the most recent supplement are a great help to the independent birder.

Finlay, J. C., ed. *Bird Finding Guide to Canada.* Edmonton, Alberta: Hurtig Publishers, Ltd., 1984. Some line drawings and simple maps; references to best times of year for special species. Good information on finding birds throughout Canada.

Lane, James A. *A Birder's Guide to* . . . Revisions to most recent editions by Harold R. Holt. L & P Press, P.O. Box 21604, Denver, CO 80221.

Southeastern Arizona (1988)
Churchill, Manitoba (1988)
Colorado (1988)
Florida (1989)
North Dakota (1979)
Rio Grande Valley of Texas (1988)
Southern California (1985)
Texas Coast (1988)

Charts show seasonal distribution in different habitats. Detailed guides to specific locations generally accompanied by local maps. Information on accommodations often included.

Pettingill, Sewell, Jr. *A Guide to Bird Finding East of the Mississippi.* New York: Oxford University Press, 1977. Detailed state-by-state coverage of good birding spots often with information on seasonal distribution.

— *West of the Mississippi.* 1981. These guides may be difficult to find but contain much useful information if used with caution.

CHECKLISTS

Listers likely will have more than one printed checklist. For many, the North American checklist constitutes their "life list." Even so, it is useful to have one of the world bird checklists. Looking through one might even tickle a world birding fancy. North American checklists are based on the most recent edition of the *A.O.U. Check-list of North American Birds* prepared by a committee of the American Ornithologists Union. The current edition is the sixth, published in 1983. It is used primarily as a reference work.

American Birding Association, Inc. *A.B.A. Checklist,* 3d ed. ABA Sales, P.O. Box 6599, Colorado Springs, CO 80934, 1986. Booklet-size for birds of North America listed in taxonomic order. Updated periodically. The ABA also publishes a small, pocket-sized *Traveler's List and Check List* giving common names only. Has multiple columns for multiple travels and an alphabetical index of common names.

Clements, James. *Birds of the World: A Checklist,* 3d ed. New York: Facts On File, Inc. 1981. (4th edition soon to be published by Cornell University Press.) Widely used by North American birders. Includes list of orders and families, bibliography, and index of scientific and common names. Checklist includes for each species a column for date of sighting, blank line for identifying the location of sightings, and one line explanation of species distribution. Book is coded for a computer system.

DeSante, David, and Peter Pyle. *Distributional Checklist of North American Birds.* Vol. 1, *United States and Canada.* Lee Vining, CA.: Artemisia Press, 1986. Large book, master checklist in taxonomic order, state and province lists, all coded for distributional status. Beautifully illustrated with full page black-and-white drawings by F.P. Bennett, Jr. and Keith Hanson. Ideal for lusty listers.

Edwards, Ernest P. *Coded Workbook of Birds of the World,* 2d ed. Vol. 1, *Non-passerines,* 1982. Vol. 2, *Passerines,* 1986. Sweet Briar, VA: Ernest P. Edwards. Loose-leaf notebook format; taxonomic list.

Howard, Richard, and Alick Moore. *A Complete Checklist of the Birds of the World.* London: MacMillan, 1984. Systematically arranged by bird families, fully indexed with scientific and English names by species. This book's particular value is that it includes every subspecies. (It is surprising how many subspecies of Northern Cardinal there are. See one cardinal and you *haven't* seen them all.)

Swift, Byron. *Checklist of the Birds of North America.* Including Middle America, the West Indies, and Hawaii. Austin: American Birding Association, 1986. (Available through ABA Sales, P.O. Box 6599, Colorado Springs, CO 80934.) This handy list correlates well with the definition of North America used in this book.

Many checklists, ranging in size from large volumes down to a size suitable for field use, are carried by book dealers listed later in this section.

GENERAL

In addition to having a field guide to help identify the bird when it is pecking about in the sand in front of you, and a checklist in which to record your sightings, most birders find it useful to have one or more general reference books to enhance their knowledge about the birds.

Austin, Oliver L., Jr. *Families of Birds: A Guide to Bird Classification.* New York: Golden Press, 1985. Useful little book, can be carried in your handbag.

—*A Dictionary of Birds.* Vermillion, SD: Buteo Books, 1985. (Campbell, Bruce, and Elizabeth Lack, ed.) Excellent reference work prepared under auspices of the British Ornithologists' Union.

Choate, Ernest A. *The Dictionary of American Bird Names.* rev. ed. by Dr. Raymond A. Paynter. Boston: The Harvard Common Press,

1985. Revised edition to accommodate the AOU 1983 checklist. Fun to look up origin of bird's name.

Connor, Jack. *The Complete Birder, A Guide to Better Birding.* Boston: Houghton Mifflin Co., 1988. Good, easy to read guide containing chapters on migration, winter, summer, optics, and acoustics. Chapters on groups of birds, such as warblers, and a useful bibliography.

Ehrlich, Paul R., David S. Dobkin, and Darryl Wheye. *The Birder's Handbook: A Field Guide to the Natural History of North American Birds.* New York: Simon & Schuster, Inc., 1988. Virtually all you ever wanted to know in a hurry about all regularly occurring breeding species. Provides information about breeding, diet, conservation, and habitat, with coded illustrations. Species descriptions are on the left-hand pages and short essays on useful information (e.g., molting) are on the right-hand pages.

Harrison, George H. *Roger Tory Peterson's Dozen Birding Hot Spots.* New York: Simon & Schuster, 1976. Describes author's birding experiences in the places identified by Roger Tory Peterson. For the curious, they are: the Everglades, south Texas, the Platte, southeast Arizona, Point Pelee, Bear River, coast of Maine, Gaspe, Hawk Mountain, Cape May (NJ), Horicon (WI), and Tule-Klamath-Malheur. Not only is this book out of print, but one of the "hot spots," Bear River in Utah, is under water. Still a good, easy to read "trip."

Kress, Stephen W. *The Audubon Society Handbook for Birders.* New York: Scribners, 1981. Great book for the early birder. Includes chapters on binoculars, spotting scopes, and photography. Reference sections include educational and research programs, periodicals (including a state-by-state listing), bird books, bird clubs, book dealers, and so on.

Leahy, Christopher. *The Birdwatcher's Companion: An Encyclopedic Handbook of North American Birdlife.* New York: Bonanza Books, 1982. Arranged alphabetically from "Aberrant" to "Zygodactyl." Includes a good topographical sketch of a bird; a taxonomic listing of the birds of North America by order, family, genus, and species; a list of North American vagrants and where they have been found; a list of migrant specialties and dates; and an extensive bibliography.

Mead, Chris. *Bird Migration.* New York: Facts On File Publications, 1983. Originally published in Great Britain, contains basic information about migration patterns and behavior worldwide. Numerous maps showing migration routes are useful.

National Geographic Society. "Bird Migration in the Americas." Washington, DC 20036, 1979. Map available for purchase.

Perrins, Dr. Christopher M., and Dr. Alex L. A. Middleton, eds. *The Encyclopedia of Birds.* New York: Facts on File Publications, 1985. Descriptions of the families of birds of the world listed in taxonomic order. Contains a glossary and extensive index. Excellent color photographs.

Rickert, John E., Sr., comp. ed. *A Guide to North American Bird Clubs.* Avian Publications, Inc. P.O. Box 310, Elizabethtown, KY 42701, 1978. Lists names of local birders who may be able to provide guidance to the visiting birder.

Root, Terry. *Atlas of Wintering North American Birds: An Analysis of Christmas Bird County Data.* Chicago: The University of Chicago Press, 1988. Analysis covers data from 1962–63 to 1971–72 in the form of relative density maps covering the major portion of the continent. Over 600 species are included with explanations of occurrence and habitat accompanying the maps. Plastic overlay maps are included, showing contours, vegetation, temperature, state and province boundaries, and national wildlife refuges.

Terres, John K. *The Audubon Society Encyclopedia of North American Birds.* New York: Alfred A. Knopf, 1980. A very *big* book, truly encyclopedic and a most useful reference guide. Contains many color photographic plates.

Zimmer, Kevin J. *The Western Bird Watcher: An Introduction to Birding in the American West.* Englewood Cliffs, New Jersey: Prentice-Hall, Inc., 1985. Both generalized and specific information on where to find birds, general information about birding, brief seasonal guide, a section on western specialties, and detailed section on bird identification.

BEYOND BIRDS

Good sources of information for birders are books about national parks and wildlife refuges, and books providing background information on places one is likely to see birds. Representative ones are listed here.

—*The Audubon Society Field Guides to the Natural Places.* Northeast, Inland and Coastal, Mid-Atlantic States, New York: Pantheon Books, 1984. Provides natural history information, detailed directions, and maps to hundreds of "natural places" including many good birding places.

National Geographic Society. *A Guide to Our Federal Lands.* Washington, DC: National Geographic Society, 1984. State by state guide to national parks, forests, wildlife reserves, grasslands, and so on in the United States. Handy paperback format provides quick reference to where special birds (e.g., Bald Eagles) can be found at what times of the year.

Reader's Digest. *North American Wildlife.* Pleasantville: Reader's Digest, 1982. For birders who seek a quick, general rundown on habitat and wildlife.

Riley, Laura and William. *Guide to the National Wildlife Refuges.* Garden City: Anchor Press/Doubleday, 1979. Good information on how to get there, what to see and do, and how to get more information. Gives substantial information on birds including seasonal information.

PERIODICALS

A wide variety of periodicals is available to the traveler/birder. Most active birders will subscribe to several in order to keep up-to-date on new books, equipment, places to bird, etc.

American Birds. National Audubon Society, 950 Third Avenue, New York, NY 10022. Ornithological journal published four times yearly according to season, plus volume on results of previous year Christmas Bird Count; caters to the interest of the serious birder. Articles emphasize New World birding. Primary attention is given to detailed listing of birds sighted in North American regions including the Hawaiian Islands and the West Indies.

Audubon. National Audubon Society. Bimonthly magazine. Well written and illustrated; covers the environment generally; often has excellent articles and illustrations on various aspects of bird life.

Bird Watchers Digest. Pardson Corp., P.O. Box 110, Marietta, OH 45750. Good bimonthly general bird watchers magazine. Sponsor of an annual workshop series for improvement of identification skills.

Birder's World. Birder's World, Inc., Holland, MI 49423. High quality bimonthly magazine. Appeared on the birding scene in 1987. Covers wide range of articles relating to birds: what and where they are, how to paint and carve them, how to photograph them, and so on. Excellent photography and reproduction.

Birding. American Birding Association, Inc., P.O. Box 6599, Colorado Springs, CO 80934. Bimonthly membership journal of the American Birding Association, devoted primarily but not exclusively to North American birding. ABA also publishes *Winging It,* a monthly newsletter. Subscriptions to both publications come with membership.

The Living Bird. Laboratory of Ornithology, Cornell University, 159 Sapsucker Woods Road, Ithaca, NY 14850. Quarterly magazine containing in-depth but generally nonacademic articles for the serious birder.

Wild Bird. Fancy Publications, Inc., P.O. Box 57900, Los Angeles, CA 90057-0900. Described as, "Your Guide to Birding at its Best," this monthly magazine began publication in 1987. Aimed at helping a wide variety of bird fanciers from feeder watchers, to birding travelers, to bird photographers. Each issue features a good birding locale.

Wingtips. Bluestone Publishing, Box 226, Lansing, NY 14882. Quarterly publication "bridging the gap between amateurs and professionals." Features meeting dates, grant notices, bird sightings, book reviews and articles, and list of ornithological organizations.

Other more general periodicals such as *National Geographic, Natural History, Sierra, Smithsonian,* and *Natural History,* often contain information valuable to the inquiring birder. Such publications are available to members of the parent organization and may be available on newstands. *Seasons,* published by the Federation of Ontario Naturalists (355 Lesmill Rd., Don Mills, Ontario M38 2W8) is a fine publication available to members. The inquiring birder traveling to different parts of the continent will encounter other local journals and newsletters containing a wealth of information about birds of a particular area.

BIRDING BOOK DEALERS

It is often difficult for birders to find an adequate selection of birding books at standard outlets. Chain book stores seldom have good selections of titles geared to specialized interests although most are glad to order special titles. Book stores and sometimes newstands and nature gift shops near popular birding areas may have better selections.

Some local Audubon organizations in birdy areas may have book stores or libraries; a few listed below engage in a well-publicized mail order business. For birders, mail order is the magic answer to finding books on where to find birds and on everything known about them once you find them.

Book stores and mail order book dealers listed here specialize in books of interest to birders and often cater to others who have broader natural history interests. Current issues of birding periodicals will keep you abreast of new distributors.

American Birding Association
ABA Sales
P.O. Box 6599
Colorado Springs, CO 80934   1–800–634–7736
Wide variety of field guides, checklists and general books available about birds of North America arranged by state. Some books on foreign birding. Some audio material listed. Discounts for ABA members.

Audubon Naturalist Society Bookshop
8940 Jones Mill Road
Chevy Chase, MD 20815   (301) 652–3606

1621 Wisconsin Avenue NW
Washington, DC 20007   (202) 337–6062

Very good selection at both locations of field guides for birding as well as other natural history pursuits; arranged by region of North America. Claims best East Coast walk-in selections of international bird field guides.

Audubon Nature Shop
300 E. University Boulevard #120
Tucson, AZ 85705   (602) 629-0757
Run by the Tucson Audubon Society. Has wide selection of natural history books.

Avian Publications
236 Country Club Lane
Altoona, WI 54720   (715) 835-6806
Emphasis is on aviculture (bird care) but contains listings of many general birding books.

Avicultural Book Company
4704 Wetzel Avenue
Cleveland, OH 44109   (216) 661-5747
Catalog lists mostly bird books with a few natural history and bird care books.

Birding Book Society
P.O. Box 1999
Salem, NH 03079-1999   1-800-262-0065
Book club offering books of interest to birders and nature lovers including identification guides, field manuals, and books on ornithology.

Buteo Books
P.O. Box 481
Vermillion, SD 57069   (605) 624-4343
Extensive list of new and old books about birds and natural history.

Capra Press
P.O. Box 2068
Santa Barbara, CA 93120   (805) 966-4590
Publishes continuing line of bird books.

The Chickadee Bookstore
440 Wilchester
Houston, TX 77079   (713) 932-1408
Operated by the Houston Audubon Sociation. Offers wide selection of bird books including field guides, bird finding guides to most

North American birding spots, checklists, and general books about birds and nature studies. Some cassettes listed.

The Crow's Nest Book Shop
Laboratory of Ornithology
159 Sapsucker Woods Road
Ithaca, NY 14850   (607) 256–5057
Gift catalog contains field guides, books, and extensive audio listings. Also available are optical equipment, T-shirts, bird feeders, and the like. Members of the Laboratory receive a discount.

The Field Naturalist
P.O. Box 161
Brentwood Bay, B.C. V0S 1A0
Canada
Bird and nature books listed by series (e.g., Stokes Nature Guide Series) as well as by major region. Some emphasis on publications pertaining to western Candada. Nature categories include mammals, entomology, butterflies and moths, reptiles and amphibians, marine life, seashells, plants, nature photography, wildlife art, and regional guides.

Flora and Fauna Books
P.O. Box 3004
Seattle, WA 98114   (206) 328–5175
Large selection of new and used bird books and books on other nature subjects.

Patricia Ledlie
P.O. Box 90, Bean Road
Buckfield, ME 04220   (207) 336–2969
Catalogs listing old and new bird books and natural history books.

Los Angeles Audubon Bookstore
7377 Santa Monica Boulevard
Los Angeles, CA 90046   (213) 876–0202
Good selection of books clearly arranged by major areas of the world or by general subject. Carries some equipment and gift items.

Massachusetts Audubon Society Bookshop
South Great Road
Lincoln, MA 01773   (617) 259–9500
Large selection of guides to birding in North America and the Western Hemisphere.

Natural History Books
1025 Keokuk Street
Iowa City, IA 52240   (319) 354-9088
Specializes in out of print and rare natural history books including a good selection of ornithological books.

Natural History Books
119 Lakeview Drive
P.O. Box 1089
Lake Helen, FL 32744-1089   (904) 228-3356
Good book lists and excellent service. Specialty is checklists compiled by owner Nina Steffe for just about every place in the world.

Nature Canada Bookshop
75 Albert Street
Ottawa Canada KIP 6B1
Well-known source of nature books.

OBServ (Odear Birding Services)
P.O. Box 1161
Jamestown, NC 27282
Offers selected recent and prepublication birding books, plus t-shirts and other birding products. See listing under *Tour Operators* later in this book.

Peregrine Heritage Tours
P.O. Box 1856
Winnipeg, Manitoba R3C 3R1 (204) 944-1169
Good source of bird books with emphasis on Manitoba. See listing under *Birding Tour Operators* later in this book.

Petersen Book Company
P.O. Box 966
Davenport, IA 52805   (319) 355-7051
Catalog contains bird books listed alphabetically by author, as well as new and out of print books on wide range of natural history topics.

R J Books & Aviary
10540 SW 160 Street
Miami, FL 33157   (305) 232-3391
Ask for catalog which covers selection of current books.

Tanager Books
51 Washington Street
Dover, NH 03820   1–800–343–9444
Attractive catalog includes descriptions of current books about birds as well as dogs, horses, and other domestic animals, horticulture, poultry keeping, and so on.

BIRDER FRIENDLY ACCOMMODATIONS

Tourist accommodations in or near birding "hot spots" provide proximity to good birding. Good birding areas in well-known tourist states are generally well-served. Some motels are advertised in magazines read by birders in a special effort to attract a birding clientele. Recent issues of such publications have been culled, and advertisers were contacted and asked to supply basic information for this section. Check current issues of such publications for other accommodations.

The Nature Conservancy operates a number of fine lodges, ranches, and other accommodations of interest to guests interested in birding and enjoying the outdoors. Write for a list. (See page 220.) Some Lane *A Birder's Guide to . . .* series booklets provide information on accommodations, as do some other bird-finding guides. Also check campground guides.

North American birding poses few accommodations problems. In remote birding "hot spots," surprisingly good places abound ranging from well-equipped camping areas to luxury hotels and elegant ranches. Hotels and motels tend to be more expensive in the east, with inexpensive motels and cabins more plentiful in the west. If in doubt, you might want to be prepared to camp, or to to fix your own meals in cabins.

Alaska

Goose Cove Lodge
P.O. Box 325
Cordova, AK 99574   (907) 424–5111
Run by birders Belle and Pete Mickelson. Lodge boasts spectacular scenery. Owners believe it is best Pacific Coast spring migration birding spot. Access to Cordova by jet service from Anchorage.

Arizona

Cave Creek Ranch
Box F11
Portal, AZ 85632

Housekeeping cottages and apartments in wooded setting within minutes of Cave Creek Canyon and other prime birding locations in the Chiricahua Mountains. Hummingbird feeders abound. Limited supplies available at Portal store.

The Mile Hi Cabins and Ramsey Canyon Preserve
RR 1, Box 84
Hereford, AZ 85615   (602) 378–2785
Six housekeeping cottages available year-round in the midst of 300-acre wooded preserve in the Huachuca Mountains, 90 miles southeast of Tucson. Bookstore and gift shop on site. Many hummingbird feeders. April through August prime birding months. Bird checklist available. Owned and operated by the Nature Conservancy. Day visitors welcome from 8 AM to 5 PM; no picnicking, pets, or recreational vehicles over 20 feet.

Ramsey Canyon Inn
Box 85, Ramsey Canyon
Hereford, AZ 85615   (602) 378–3010
A Bed & Breakfast adjacent to the Mile Hi/Ramsey Canyon Preserve. This mountain inn has both private rooms and shared baths. Bird list of Ramsey Canyon and surrounding area available.

Santa Rita Lodge
P.O. Box 444
Amado, AZ 85640   (602) 625–8746
Cottages and motel units with kitchens available year-round in Madera Canyon within the Coronado National Forest 40 miles south of Tucson. Service facilities 20 minutes away. Birding good year-round but best March through June. Bird list available.

Stage Stop Inn
P.O. Box 777
Patagonia, AZ 85624   (602) 394–2211
Apartments, efficiencies, and rooms, with swimming pool and restaurant, right in the heart of Patagonia. Very near Sonoita Creek Bird Sanctuary.

British Columbia

Merlin House
3672 West First Avenue
Vancouver, B.C. V6R 1H2   (604) 736–9471

Bed & Breakfast "catering only to naturalists and birdwatchers." Convenient to Reifel Bird Sanctuary and other well-known birding areas.

Connecticut

Applewood Farms Inn
528 Colonel Ledyard Highway
Ledyard, CT 06339   (203) 536-2022
A Bed & Breakfast guest house with six rooms. Adjacent to Nature Conservancy lands and near Mystic, Connecticut.

Country Goose Bed and Breakfast
RFD #1, Box 276
Kent-Cornwall Road
Kent, CT 06757   (203) 927-4746
Guest house has four rooms. Located near the Northeast Audubon Center in northwest part of state.

Florida

Bass Haven Lodge
P.O. Box 147
Welaka, FL 32093   (904) 467-2392
Ten lodge rooms and efficiencies. Open year-round. Overlooking St. John's River, northwest of Crescent City. Caters to fishers but welcomes birders. Fishing equipment available.

Flamingo Lodge, Marina & Outpost Resort
Everglades National Park
Flamingo, FL 33030   (305) 253-2241
One hundred and two rooms in the Lodge and 16 cottages on Florida Bay near park headquarters. Restaurant, small convenience store, gift shop, tackle shop, and gas station. Winter is best birding season (very buggy the rest of year; limited services May 1 to October 31). Bird list available; also lists of mammals, amphibians, and reptiles.

Pelican Inn
P.O. Box 301
Carrabelle, FL 32322   (904) 697-2839
Seven efficiency rooms on Dog Island, 50 miles south of Tallahassee. Accessible by small plane, boat, or passenger ferry. Inquire about available limited group menus. Bird list available.

Georgia

Little St. Simons Island
P.O. Box 1078 AL
St. Simons Island, GA 31522
Overnight accommodations for 24 in three lodges, one cottage on privately owned island offshore from Brunswick reached by LSSI ferry. Dining room specializes in seafood and southern specialties. Bird guides and lists available.

Maine

The Hitchcock House
Monhegan Island, ME 04852   (207) 372–8848
Five rooms, three with kitchens, located ten miles offshore from Port Clyde. Reached by mailboat. Meals available nearby. Best time: May/June and September/October. All rooms have bird feeders.

Matinicus Island Cottages
Matinicus Island, ME
Two cottages available May through October. Access by ferry from Rockland, or air charter from Owl's Head or from Portland. Staples available on island. Boat service available to nearby Matinicus Rock and Seal Island with their Atlantic Puffin breeding colonies. Bird list available. Bookings through Geoffrey G. Katz, 156 Francestown Road, New Boston, NH 03070; (603) 487–3819.

Massachusetts

Red River Motel
Route 28
South Harwich, MA 02661   (617) 432–1474
Rooms and efficiencies open year-round, located on the south side of Cape Cod. Restaurants and shops nearby. Bird tours can be arranged through Massachusetts Audubon Wildlife Sanctuary.

Seaward Inn
Rockport, MA 01966   (617) 546–3471
Rooms and suites in the Inn and cottages open mid-May to mid-October, overlooking ocean on Cape Ann. Restaurant. Shops in nearby village. Audubon sanctuaries nearby.

Minnesota

Allyndale Motel
510 North 66th Avenue West
Duluth, MN 55807   1–800–341–8000
Motel rooms and kitchenettes convenient to expressway and downtown. Pleasant grounds with bird feeders and adjacent woods.

Bunt's Bed & Breakfast
Lake Kabetogama
Ray, MN 56669   (218) 875–3904
Lodge located on 20 forested acres adjacent to Voyageurs National Park. Three private rooms with sleeping accommodations expandable for groups up to 14. Open year-round. Bird list being prepared.

Montana

Pine Butte Guest Ranch
HC58, Box 34C
Choteau, MT 59422   (406) 466–2158
Operated by the Nature Conservancy, this ranch is located at the foot of the East Front, near the Bob Marshall Wilderness area. Accommodating 20 guests in eight "rustic, yet comfortable" cabins, the ranch is open from May through October. Transportation is provided from Great Falls International Airport.

New Brunswick

Shorecrest Lodge
North Head
Grand Manan
New Brunswick E0G 2M0   (506) 662–3216
Country inn containing fifteen guest rooms, open May to October. Access via car ferry from Black's Harbor. Family-style meals. Boat access to Machias Seal Island puffinry; good bird and whale watching. Bird list available.

New Jersey

West Cape Motel
303 Sunset Boulevard
Cape May, NJ 08204   (609) 884–4280
Hangs out a "Welcome Birders" sign. Minutes from Hawk Observatory. Twenty units, most of them efficiencies, and four apartments.

New Mexico

Bear Mountain Guest Ranch
P.O. Box 1163
Silver City, NM 88062   (505) 538-2538
Bed & Breakfast inn open year-round serving three meals daily; sack lunch for birders. Seventeen units with private baths, and two cottages. Good birding on property and in nearby preserves. Owned by well-known birder Myra B. McCormick. Birding good all year, especially during migration: mid-April to early May and late September to early October. Birding tours and bird list available.

South Carolina

Kingfisher Cottage
995 Buck Hall Landing
McClellanville, SC 29458   (803) 887-3327
Located on the intracoastal waterway, cottage has two bedrooms and sleeps four. Overlooks Cape Romain National Wildlife Refuge. With a name like that, you know owners Mike and Donna Ratledge cater to birders.

Texas

Indian Blanket Ranch
P.O. Box 206
Utopia, TX 78884   (512) 966-3525
"Bed & Breakfast for Bird-Watchers." Two rooms and cabin in the woods open year-round on 250-acre ranch on the Edwards Plateau, 90 miles west of San Antonio. "Folksy Gourmet" meals available. Owners knowledgeable birders. Birding good year-round, especially during migration. Specialty of area is Golden-cheeked Warbler that nests nearby. Bird list available.

Village Inn Motel
503 North Austin
Rockport, TX 78382   1-800-338-7539
Twenty-eight units including some kitchenettes and apartments. Provides free continental breakfast and offers help to birders.

Surfside Motel
1809 Broadway
Rockport, TX 78328   (512) 729-2348
Thirty-six units, some with kitchenettes, on saltwater bay. Caters to birders and can recommend local birding expert. Provides useful

information on area and birds including a bird list for the area surrounding the motel. Information available on boat trips to view Whooping Cranes.

Your travel agent can best advise you about accommodations in Mexico, Central America, and the islands. Particularly birder friendly accommodations are the following:

Costa Rica

Hotel La Mariposa
P.O. Box 4
Quepos, Costa Rica   77–03–55
Small luxury hotel accommodating 40 guests, located on Pacific coast amid a lovely garden setting. Open year-round.

Las Ventanas de Osa
Costa Rica
Inquiries and bookings through Natural History Books, Inc., P.O. Box 1089, Lake Helen, FL 32744-1089. Lodge accommodations located in virgin rainforest 500 feet above Pacific Ocean, open January through April. Reached by four-wheel vehicles from San José. Six double rooms, four single rooms; restaurant, bar, and swimming pool. English-speaking staff. Birding tours and bird list available.

TOUR OPERATORS

There is no infallible answer to the question, "Should I go birding on my own or with a group?" Both styles have advantages. If you do decide to go with an organized group, there is another set of decisions to be made. Do you want to go with a group whose primary focus is birding? Such groups provide you with a bird checklist, make every effort to help you see all the species in a particular area, and keep a tight schedule.

Other tour operators provide marvelous opportunities to see birds, but aren't solely dedicated to birding. Nature oriented tour operators provide a more broadbased experience and more opportunities for absorbing the atmosphere and culture of an area (although this will not be entirely lacking on a birding tour). Such tours generally are more leisurely than strictly birding tours. Some tour operators will take you to the same places birding tours go, but are not particularly oriented to bird life.

Pelagic birding tours are yet another story. Sighting ocean birds may be possible anytime you are out on the ocean and away from the shore. It may be the Bluenose ferry ride between Bar Harbor, Maine, and Yarmouth, Nova Scotia, or a chartered fishing boat going out into the Gulf Stream. There are many opportunities off both coasts for combining whale watching with sea birding. The same thing may hold for snorkeling and diving trips. Some bird tour operators offer both land and sea birding.

In listing some of the better known tour operators, the strictly birding tours are listed first. The distinction between birding tours and nature tours is far from precise. Readers interested in rigorous birding trips, or birding/natural history trips, are urged to review the catalogs of both. Look also in recent publications read by birders for other tour operators going to places you want to go. Many offer trips beyond North America, but only North American destinations in the context of this book are listed.

Not all group birding experiences need be with specialized tour operators. Some local or state Audubon Societies offer trips to special birding locations both nearby and faraway. For information on group workshops or local trips, check the Audubon societies in areas identified in this book.

BIRDING TOUR OPERATORS

Some birding tour operators specialize in particular areas. Others offer a wide variety of destinations. Some trips have waiting lists, while others are canceled because of lack of interest. Inquire early, but don't hesitate to call at the last minute to see if there are cancellations.

Attour, Inc.
P.O. Box 1353
Highland Park, IL 60035   (312) 831–0207
Originated and organized by Larry Balch, this is *the* way to have an Attu birding experience.

Bird Bonanzas, Inc.
P.O. Box 611563
North Miami, FL 33161   (305) 895–0607
Led by Joel Abramson, identified by the American Birding Association as one of the top world birders, advertises economical tours featuring expert bird leaders. The 1988 schedule included weekend trips to Texas, New Mexico, south Florida, and several offshore areas, especially the West Indies.

Budget Birding
1731 Hatcher Crescent
Ann Arbor, MI 48103   (313) 995-4357
A relative newcomer on the bird tour scene, this operator offers trips at a price excluding food. Their 1989 schedule includes winter birding in Michigan's Upper Peninsula, spring migration at Point Pelee and Southern Ontario, spring raptoring at Whitefish Point, Michigan, and fall raptoring at Holiday Beach, Ontario.

Cardinal Birding Tours
P.O. Box 7495
Alexandria, VA 22307   (703) 360-4183
Led by Don Peterson, Howard Langridge, and Erika Wilson; concentrates on south Florida and Chesapeake Bay area, Texas, and California. Emphasizes one-week, small-group tours.

Cornell Laboratory of Ornithology
159 Sapsucker Woods Road
Ithaca, NY 14850   (607) 255-3341
Sponsors limited number of birding tours led by ornithologists and utilizing other tour operators.

Field Guides, Inc.
P.O. Box 160723
Austin, TX 78746   (512) 327-4946
Led by Jan Erik Pierson, John Rowlett, Rose Ann Rowlett, Bret Whitney, John Arvin, Chris Benesh, John Coons, Tom Crabtree, Doug McRae, Frank Oatman, Wayne Peterson, and Dave Stejskal; the 1988–89 program included many North American birding "hot spots," along with several in Middle America and the West Indies.

Four Points Nature Tours, Inc.
P.O. Box 23825
Chattanooga, TN 37422   (615) 757-1999
Led by Benton Basham and Noble Proctor, offers land and pelagic birding trips to Alaska, Arizona, Dry Tortugas, Costa Rica, and other destinations.

OBServ (Odear Birding Services)
P.O. Box 1161
Jamestown, NC 27282
Offers trips led by Bob Odear and other expert birders to selected North America birding "hot spots," and pelagic trips off North Carolina. See listing under *Birding Book Dealers* earlier in this Appendix.

Massachusetts Audubon Society
Lincoln, MA 01773   (617) 259–9500
Sponsors birding and nature tours led by experts. The 1988–89 schedule included Alaska, Baja, Florida, Churchill, and the southwest United States.

McHugh Ornithology Tours (MOT)
101 West Upland Road
Ithaca, NY 14850   (607) 257–7829
Provides several North American tours led by experts. Some trips sponsored by Cornell University Laboratory of Ornithology.

NatureAlaska Tours
P.O. Box 10224
Fairbanks, AK 99710
Led by Dan L. Wetzel; focuses on birding and nature study in his home state, particularly arctic Alaska.

Parula Tours
1711 West Oglethorpe Avenue
Albany, GA 31707   (912) 439–8232
Small group tours to Florida, Alabama, Georgia, Texas, and Arizona.

Peregrine, Inc.
1521 West St. Mary's Road
P.O. Box 250
Tucson, AZ 85745   (602) 823–4295
Led by Ben Feltner, with associates Dan Wetzel, Hilary Thompson, Linda M. Feltner, and Dr. David Mark. The 1987 schedule featured the Pacific Northwest, Texas, Southeast Arizona, and Alaska. Also provides "short notice" trips to see special species, private guide service, and field identification seminars for groups.

Peregrine Heritage Tours
P.O. Box 1856
Winnipeg, Manitoba R3C 3R1   (204) 944–1169
Wayne Neily and other knowledgeable bird leaders offer both general and customized bird trips in this bird-rich area. See listing under *Birding Book Dealers* earlier in this Appendix.

Raptours
P.O. Box 8008
Silver Spring, MD 20907   (301) 565–9196

Raptor workshops and bird tours specializing in hawks and owls, led by Bill Clark, author of the Peterson guide, *Hawks;* Brian Wheeler, artist for that book; and other knowledgeable leaders.

Victor Emanuel Nature Tours (VENT)
P.O. Box 33008
Austin, TX 78764   1–800–328–VENT
Led by Victor Emanuel, David Bishop, Steve Hilty, Ted Parker, Bob Ridgely, David Wolf, Kevin and Barry Zimmer, Dale Delaney, and Ken Kaufman. The 1989 tour schedule included Alaska, Arizona, Texas, Florida, Churchill, Mexico, the West Indies, and other North American "hot spots." Weekend and photographic trips are also offered.

Wings, Inc.
P.O. Box 31930
Tucson, AZ 85751   (602) 749–1967
Led by Will Russell, Davis Finch, and other expert leaders. Offers many trips to "birdy" places in North America, Mexico, and the Caribbean.

WoodStar Tours
908 South Massachusetts Avenue
DeLand FL 32723   (904) 736–0327
Offers birding trips to Costa Rica.

PELAGIC BIRDING TRIP OPERATORS

These tour operators are known for their emphasis on pelagic birding. For other tour operators offering opportunities for oceanic birding, see the *Nature Tour Operators* listing which follows. Whale watching tours are also a good way to combine seabird watching with watching whales, porpoises, and other creatures of the ocean.

Paul G. Dumont
750 South Dickerson Street Apt. 313
Arlington, VA 22204   (703) 931–8994
Offers pelagic trips off the North Carolina coast from July through September.

Shearwater Journeys
c/o Sam's Fishing Fleet,Inc.
84 Fisherman's Wharf
Monterey, CA 93940   (408) 688–1990

Led by Debra Love Shearwater; utilizes additional pelagic birding experts. Offers extensive year-round calendar of one-day and several-day trips for Pacific Ocean birds. Most frequent trips: August–October.

T. R. Wahl
3041 Eldridge
Bellingham, WA 98225   (206) 733–8255
Pelagic boating trips offered out of Westport, Washington; winter seabird trips out of Anacortes, Washington.

Pelagic trips are often offered by Audubon Societies and bird clubs in coastal areas. Contact organizations directly for information on ocean trips out of major coastal cities.

NATURE TOUR OPERATORS

Listing includes some seagoing expeditions that may provide opportunity for birding.

Alaska Fishing and Wilderness Adventures
P.O. Box 102675
Anchorage, AK 99510-2675   1–800–544–2219
Natural history safaris and cruises. Opportunities for pelagic birding.

Biological Journeys
1876 Ocean Drive   (707) 839–0178
McKinleyville, CA 95521   1–800–548–4555
Led by Ronn Storro-Patterson, naturalist and marine biologist; Ron Levalley, naturalist and bird expert; and John Kipping, naturalist. General nature tours with whale watching and birding orientation. The 1988 program offered trips aboard the Delphinus, sleeping ten passengers, for whale watching and pelagic birding in Baja California, the Pacific Northwest, Alaska, and the Galapagos. One-day pelagic trips operated out of San Francisco.

Canadian Nature Tours
Fon Conservation Centre
355 LesMill Road
Don Mills, Ontario M3B 2W8   (416) 444–8419
Birding and nature tours available to members of the Canadian Nature Federation and the Federation of Ontario Naturalists.

Cheesemans' Ecology Safaris, Inc.
20800 Kittredge Road   (408) 741-5330
Saratoga, CA 95070   (408) 867-1371
Led by Doug Cheeseman, zoologist and ecologist, and Gail
Cheeseman, naturalist and birder; emphasis on birding, natural
history, and wildlife photography. Participants must be nonsmokers.
The 1988 program included Monterey Bay, California, whale watch-
ing.

Churchill Wilderness Encounters
P.O. Box 9-B
Churchill, Manitoba   (204) 222-7877
Canada R0B 0E0   (204) 675-2248
Birding and general nature tours during the summer; polar bear
trips in the fall.

C.V.S. Inc.
Natural History Expeditions
P.O. Box 1144
East Dennis, MA 02641
Birding and natural history sailing trips. The 1988–89 sched-
ule included Georgia Sea Islands, Carolina Low Country, and Nova
Scotia.

Earthwatch
P.O. Box 403
Watertown, MA 02272   (617) 926-8200
A unique organization providing field research opportunities in a
number of disciplines around the world. Although the research is unre-
lated to birding, a particular place might provide opportunities for the
birder. Several programs are led by ornithologists engaged in research
projects.

Flip Pallot
5555 SW 67th Avenue #107
Miami, FL 33155   (305) 245-3800
Specializes in group trips through Florida Everglades.

Holbrook Travel, Inc.
3520 NW 13th Street
Gainesville, FL 32601   (904) 377-7111
General natural history and cultural trips, including some especially designed for birders and led by birding specialists.

Joseph Van Os Nature Tours
P.O. Box 655
Vashon Island, WA 98070   (206) 463–5383
Led by Joe Van Os and other expert leaders for particular areas; emphasizes general interest in nature study and photography, especially birds. The 1988 schedule included the high Arctic, Yellowstone, the Grand Tetons, Washington state, British Columbia, Florida Everglades, Arizona, Alaska, and polar bear trips to Churchill, Manitoba.

Key West Seaplane Service
5603 Jr. College Road
Key West, FL 33040   (305) 294–6978
Provides "adventure flights" to the Dry Tortugas with half day, full day, and camping trips. Reservations required.

Nature Expeditions International
P.O. Box 11496
Eugene, OR 97440   (503) 484–6529
Natural history tours with some emphasis on bird life, includes trips to Alaska, Hawaii, Mexico, and Baja California.

Nature Travel Service
127A Princess Street
Kingston, Ontario
Canada   (613) 546–3065
A wide variety of birding and nature tours are offered for destinations in Canada, the United States, Mexico, and Central America. President Gus Jaki has attracted a large and devoted following.

Oceanic Society Expeditions
Fort Mason Center, Building E
San Francisco, CA 94123   (415) 411–1106
Land and ocean trips led by naturalists, with emphasis on birding and whale watching. Trips include Baja California and the Sea of Cortez, Vancouver Island, the San Juan and Gulf Islands, Queen Charlotte Islands, southeast Alaska, Bahamas, Hawaii, and Bay of Fundy (with local boat trips for whale watching and island birding).

Pacific Sea Fari Tours
2803 Emerson Street
San Diego, CA 92106
Focus on Baja California whale watching but also providing opportunities for ocean birding. Bird list available. Charter trips available.

Questers Tours and Travel, Inc.
257 Park Avenue South
New York, NY 10010-7369   (212) 673-3120
Worldwide nature tours with strong birding orientation.

Society Expeditions, Inc.
723 Broadway East
Seattle, WA 98102   1-800-426-7794
Designs and markets worldwide sailing of luxury expedition ships:
the *World Discoverer* and *Society Explorer*. The 1988 schedule included
several cruises to Northern Canada, Alaska and islands in the Bering Sea.

Special Odysseys
P.O. Box 37
Medina, WA 98039   (206) 455-1960
Specializes in the Canadian Arctic and Greenland. Offers a number
of trips of particular interest to birders. Special individual and group
itineraries can be arranged.

World Nature Tours, Inc.
P.O. Box 693, Woodmoor Station
Silver Spring, MD 20901   (301) 593-2522
Offers nature tours with emphasis on birding. The 1988 destina-
tions included several Texas birding "hot spots," southeastern Arizona,
Washington, and Oregon. Society Expeditions, Inc.

National Audubon Societies and various state and local Audubon groups
offer birding and nature tours, some of which are led by ornithologists
and other knowledgeable birders. Many other organizations such as the
Smithsonian Institution and Sierra Club, offer great nature trips which
may provide opportunities for birding without the intensity of a typical
birding trip.

THE WELL-EQUIPPED BIRDER

A swimmer without a bathing suit? Well, now! A tennis player without a
racket and balls, or maybe even a guide to better tennis? A camper
without a tent, stakes, sleeping bags, cookstove, and so on? (Did you
ever leave the center pole back home?) For birders it is no different.
Doubtless you already have the two basic pieces of birding "equipment": binoculars and field guide. You may, however, be interested in
upgrading your binoculars, or perhaps you don't really like the scope

your husband/wife gave you for Christmas. Maybe you are ready to tape some bird calls. You can keep the sport of birding as simple as cane pole fishing, or you can arm yourself with some pretty sophisticated equipment.

BINOCULARS

No self-respecting birder will be caught without binoculars anytime, anywhere. Avid birders who can see out a window without getting out of bed in the morning have been known to keep binoculars on the night table. When you decide to upgrade, keep the old ones under the seat of the car. While waiting for a traffic signal to turn green (if you are in south Florida), you might get a good look at a Common Myna among the European Starlings sitting on the wire. Binoculars can also be turned upside down and used as a microscope. Try it the next time you want a good look at a minute tundra flower.

Although many early birders worry about what binoculars to buy, there are many that will suit the purpose. The first binoculars bought are often 7×35 or 8×40. These are useful for most birding; in fact some birders prefer them. A "quick-and-dirty" survey a few years ago of birders on a trip to the Dry Tortugas in Florida indicated that when birders upgraded, the favorite glasses were Zeiss or Leitz 10×40's.

Briefly, the numbers mean two things: magnification and brightness of image. The first number measures the magnification: a bird will look 7 times or 10 times as close as it is. The second number measures the amount of light that comes through the objective lens, the large end of the binoculars. The higher the number, the more light. Thus the number 40 indicates that more light comes in than the number 35; the image will be brighter. There's more to it than that. Veteran birder Peter Alden adds a few pointed remarks:

> Everything else being equal as you narrow down your choice of binoculars, you should then compare minimum focal distance. To this day no one can figure out why so many binoculars are designed that focus down to only 25 feet, but at the other extreme you can focus on to the moon, and way beyond infinity. Whether you're birding in the gloom of Amazonia, or just pished a fall warbler into a nearby thicket, there are abundant times when birds come so close, you have to move back to get it in focus. Some models allow you to focus down to 8 feet or so, while others can be factory adjusted for minimal focus distance. Wide fields of view are not terribly important except to rank beginners. Like shooting a rifle, or murder, it gets easier and easier with a little practice, and you'll soon find that when a bird pops up, you'll automatically zero in instantly.

Experts agree: buy the best you can afford. But read an article or two to get an idea of what you will be comfortable with. Get on the mailing list of one of the mail order book dealers to keep current with new releases. Binocular manufacturers in recent years have begun to listen to expert birders about what they want in binoculars, not for checking on some ship way out at sea or watching a football game but for birding. New models appear periodically and most of the birding magazines request birding experts to try them out and report on them. A major evaluation and rating chart, "Scanning for Optics" by Richard E. Bonney, Jr. and Jill Baringer, appeared in *The Living Bird Quarterly,* publication of the Cornell Laboratory of Ornithology, the Autumn 1988 issue.

Binoculars can be purchased at local outlets such as department stores and camera shops, or they can be mail ordered from organizations specializing in birding books and birding equipment.

SCOPES

Not quite as much information is available about scopes as about binoculars although the birder who has recently purchased one will likely wonder how a birder can cope without a scope. Lacking one, you may miss some "good" ducks, geese, and shorebirds way out in the marshes and on the sandbars. That raptor perched on the snag just misses with binoculars, but is clearly a Peregrine as seen through the scope. If you do much birding from your car, you may want to get a scope mount that will fit on the car window. Most birds that are skittish about people are not disturbed by automobiles. In fact, the car makes a good blind for scoping or photographing.

Although open areas such as mud flats, marshes, or prairies are obvious places where a scope is needed, world bird leader Alden suggests, "Telescopes can be of great use in open or dense forests as many a hawk, trogon, motmot, hornbill, or barbet does sit quietly for extended periods. Forest birds are more shy than grassland birds, and the telescope can be zeroed in and allow your party to all see the bird without loud, near hopeless descriptions of hundreds of look-alike limbs and leaves." Peter Alden's long arms always seem to have a scope and tripod attached. On the subject of tripods, he advises that the one with the fewest gadgets to turn is the best.

If you go with a birding tour group, your leader will likely have a Questar. "I've got it (the bird everyone is trying to find) in the Quester!" That's your leader's signal to quickly line up for a fine view of the Acadian Flycatcher. Questars are expensive scopes, but if that factor doesn't bother you, do consider it.

Some published information about scopes is included in articles on

Birders in Churchill, Manitoba, use scopes for a good look at rare Ross's Gulls

binoculars. A good comprehensive review of scopes by Tim Gallagher appeared in the August 1988 issue of *Wild Bird.*

TAPES AND RECORDS

Many birds that sing so sweetly, are shy, retiring and difficult to see. Vireos generally would fit into this category. It helps enormously to be able to identify such birds by their song or call. Listening to records or tapes of North American birds is a good idea, particularly if you plan a visit to a part of the country that has species different from the ones you know well. If you have a tape deck in the car, you can entertain yourself in rush hour traffic instead of fuming at the snail's pace of your progress.

A second use for tapes involves a portable tape recorder, small enough to fit into a jacket pocket. It's good for quickly recording information about the birds you are seeing. Using a good tape recorder with a "shotgun" microphone enables recording the song of some evasive bird, then playing it back in hopes of enticing the skulker out of the dense shrubbery. Some birders will make their own tape recording of species they are likely to see, or of calls that will attract other birds. The Lane guide to southeast Arizona suggests a tape of the Pygmy or Whiskered Owl as one to use. Be respectful, however, of local regulations that may ban the use of such tapes. At nesting time, taped calls may seriously disrupt an egg sitter's life.

Numbers of bird call tapes available today are beginning to approach the number of books about birds. Some book dealers listed previously also carry audio and video materials. They will be glad to get your inquiry. Special mention needs to be made of the Cornell Laboratory of Ornithology. This organization is a pioneer in the taping of bird calls and has a fine collection. See "The Crow's Nest Bookshop" listed under *Birding Book Dealers* earlier in this book. A good review of recordings can be found in a *Birding* magazine, February 1986, article by William M. Meriwether entitled "ABA-area Bird-sound Recordings."

Audio aids keyed especially to two of the widely used field guides are:

A Field Guide to the Bird Songs of Eastern and Central North America, 2d ed. Metromedia Producers Corp. and Houghton Mifflin Co., Boston. Available from Cornell Laboratory of Ornithology, Ithaca, NY 14850. Discs or cassettes. Designed to accompany Roger Tory Peterson's *Field Guide to the Birds East of the Rockies.*

A Field Guide to Western Bird Songs. To be used with *A Field Guide to Western Birds.* Tapes or records.

Guide to Bird Sounds. National Geographic Society, Washington, DC. Cassettes. To be used with National Geographic Society *Field Guide to the Birds of North America.*

VIDEOTAPES

Inevitably, we use the television to learn about birds and, within the last several years, videocassettes have been used to improve birding skills. These cassettes tend to be relatively expensive so you might want to persuade your bird club to purchase and lend them to members.

Audubon Society's VideoGuide to the Birds of North America. MasterVision, 969 Park Ave., New York, NY 10028, 1988. A five-cassette series prepared by Roger Tory Peterson and Michael Godfrey.

Gone Birding! The VCR Game. Novel 1989 video board bird identification game with Peter Alden, Bill Oddie (British birding "personality"), and Jane Alexander. Rupicola VCR Games, Inc., 1300 Washington Street, Walpole, MA 02081. Could it become as popular as Monopoly? It looks like a fun way to learn.

Techniques of Birding with Arnold Small. Nature Videos, P.O. Box 312, South Laguna, CA 92677.

Watching Birds with Roger Tory Peterson. Metromedia Producers Corp. and Houghton Mifflin Co., Boston.

COMPUTERS

It seems that everything is computerized these days, and birding is no exception. Current birding periodicals advertise a variety of programs to help keep your bird lists in order. Most are available from your favorite mail order book dealer. The Clements world bird checklist is computer coded.

BirdBase. Santa Barbara Software Products, 1400 Dover Road, Santa Barbara, CA 93103, (805) 963–4886. Data base of 721 North American species; accommodates addition of accidentals, splits, and lumps; up to nine different lists. Compatible with *WORLD BirdBase* utilizing 1989 update of Clements *Birds of the World: A Checklist.* Demo disk available for latter.

Bird list. Bird Commander, P.O. Box 34238, Bethesda, MD 20817, (301) 229–7002. Prepared by Tony White; a North American life and annual list to help keep track of your sightings. Computer program includes Hawaii. Publishes *Birdlist Quarterly Newsletter* full of interesting guidance to birding hot spots as well as updated information for the computer program; and annual *Distribution of North American Birds* identifying species by states, provinces and territories.

Plover and *Gecko.* Sandpiper Software, 153 Michele Circle, Novato, CA 94947. Programs containing information to use in keeping your list up to date. Check with them for specifications on the most recent version.

Further information on birding bookkeeping and how to adapt software for birding purposes can be gleaned from Edward M. Mair's *A Field Guide to Personal Computers for Bird Watchers and Other Naturalists.* Englewood Cliffs: Prentice-Hall, 1985. Like everything else in the computer world, books such as this become quickly dated, but you may pick up some useful hints.

MAPS

Birders who travel and bird on their own know the importance of good maps. The most detailed highway map you can find is the best. It will show rest stops, parks, and wildlife areas, plus country roads and how they connect back to the interstate. If you plan to spend time at wildlife refuges in remote areas, be sure to ask for their detailed map. At places like Pawnee National Grasslands in northern Colorado, roads may be hard to find and often are unmarked. You may not see another soul for hours so there's nobody to tell you to go straight ahead on the rutted road instead of turning right on the better road. It really helps to have a detailed map showing every barn, windmill, and water tower if you hope to see the McCown's Longspur where your Lane guide says it is. Wildlife refuges, in addition to national parks and forests, often have such detailed maps.

ABA members have available to them the *Birdfinding Guide,* a loose-leaf notebook providing detailed information, usually including maps, about locations for finding particular species. Updated inserts available.

INVISIBLE EQUIPMENT

This may, unfortunately, be the scope you left on the tripod at the mudflats back there—and it was gone when you returned. The invisible equipment we are talking about is what birders bring with them into the field. Good physical condition, alert eyes, and trained ears atuned to a distant bird's song or a near "twit"—these are the kinds of personal equipment important to good birding. American birders generally don't listen as well as their British counterparts. On an April buggy ride to see the rails at the Anahuac National Wildlife Refuge, British birders were "seeing" birds the rest of us were not. They noted "something" fly off and nonchalantly identified "Swamp Sparrow." They identified its trilling song. "It saves wasting time chasing birds you've seen dozens of times. That call over there—*that's* different."

RARE BIRD ALERTS

If you are an early birder, you may not care to know about what rare birds have invaded your home territory, much less the continent. News that a Sharp-tailed Sandpiper has been seen 3000 miles from where you live may elicit no greater response than a quiet yawn. Your attention is concentrated on identifying birds. As your birding list grows, however, you may well want to be alerted that a bird you haven't yet seen is in the neighborhood.

Many local bird clubs have either formal or informal means of alerting members about special birds that have flown into the territory. Some state birding organizations offer hotlines. The American Birding Association publishes periodic listings of the telephone numbers. Some local groups have listed hotlines, so check the telephone directory in the city you are visiting. The south Florida birder visiting Colorado may not think that the Denver Field Ornithologists hotline report of a Purple Gallinule is anything to get excited about, but lots of Colorado birders have never seen one. On the other hand, a birder seeking current news from the San Francisco Audubon Society opted to stay at a particular motel because of reported sightings of special species nearby.

Eager birders flushed with the success of identifying 500 North American species and impatient to knock off 600 (and then 700!), use the national bird alert. Subscribers to the North American Rare Bird Alert (NARBA) have several means available of finding out what birds are being seen *today* in different parts of the continent. NARBA keeps current a list of state rare bird hotlines. Write or call for further information:

North American Rare Bird Alert
c/o Houston Audubon Society
440 Wilchester
Houston, TX 77079
Canada   1–800–438–6704
U. S.   1–800–438–7539

PHOTOGRAPHY

After visiting Everglades National Park in February, California early birder Bert Bream was seriously considering shifting his focus from watching birds to shooting them with his camera. He couldn't get over the thrill of having a Great Egret absolutely fill the frame of his camera. And they stand so still.

Some birders are superb photographers. Some bird photographers are excellent birders. The two activities, however, are difficult to combine unless you are birding alone, in places of your own choosing, and have unlimited time. On an organized birding trip, the group is interested in seeing as many species as possible in an all too little time. Photography must often be done "on the fly." Being serious about photography means adding more heavy equipment. Many birders find they just don't have as many hands or as strong a back as they need.

Attempting to photograph birds often results in a large dose of frustration. Most forest birds of the tropics are high in the canopy busily feeding, flitting from branch to branch. In the wetlands, the Least Bittern you would most like to photograph is hidden in the grasses on the *other* side of the pond. One birder-photographer commented, "I have lots of slides showing habitat, but not many showing birds." The one day you see the Ross's Gull the sky is overcast and

Birding photo takers like large birds (above),

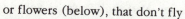

or flowers (below), that don't fly

your slide just misses the delicate pink on its breast. Your companions are losing patience with you; the click of the shutter scares off the bird they were studying.

On the other hand, for many of us, photography is a natural adjunct to birding and we sling a camera over our shoulder any time we're birding. Getting a reference shot of a rare bird may prove you saw it. Some birders, like Maida Maxham, try to get a photograph of each species seen. She comments, "Most birds have not been overly cooperative; nonetheless I press on, striving to capture in film the glint in the eye, the glorious detail of feather, habitat, and behavior that defines each bird." Photographs of discernible birds and the general birding activity, can bring back fond memories of a birding adventure and may produce a good slide show for the local bird club. Photographs of "dot" birds may also recall a good birding trip even though only you know that there is a Clay-colored Sparrow in that otherwise featureless bush.

The field of photography has its own library of books and periodicals. Birders interested in bird photography would do well to look at the bimonthly *Birder's World* for some excellent examples of bird photography and often helpful articles.

GENERAL EDUCATION

For the birder who wants to know more about birds, their characteristics, their behavior, their flight, their nests, and so on, there may be classes in ornithology at the local university. If that is not a good option, the Laboratory of Ornithology at Cornell University has a home study course in bird biology which provides an excellent substitute. Request information about the "Seminars in Ornithology" from:

Laboratory of Ornithology
Cornell University
159 Sapsucker Woods Road
Ithaca, NY 14850

Additionally, several organizations offer workshops in some aspect of birding. The National Audubon Society, *Bird Watcher's Digest,* and some local and state birding organizations schedule them from time to time. Some universities schedule occasional workshops. One of the better known is the University of Maine at Machias, Institute for Field Ornithology (9 O'Brien Avenue, Machias, ME 04654).

COMMON SENSE

Every sport has its hazards. Not many for birders who have confined birding to the backyard. Birding travel does have its perils, but some general precautions and old fashioned common sense protect all but the most foolhardy. Many common sensities apply to places beyond the beaten path and some special ones apply to birders their special habitats. Many birding travelers have contributed to this list; you can add notes of your own.

VALUABLES

Avoid leaving a purse in a locked car where it can be easily spotted. Preferably don't take one along. Ratty old canvas bags generally don't invite unwanted curiosity. Think about leaving good watches, gold jewelry, and diamond rings at home. Don't invite problems.

You need not keep your "powder dry," but do *keep your binoculars and cameras dry*. Birders do bird in the rain, and a rain guard for binoculars is a good idea. Holding them under your arm or tucking them inside your jacket can help. A plastic bag for equipment not in immediate use will not only keep it dry, particularly important on ocean trips, and will keep out fine dust from rough country roads in the dry season.

HEALTH PRECAUTIONS

In tropical areas, *don't drink the water* unless you are told it's safe. However, *do* drink lots of liquids to help keep you on a regular schedule. If you know the water isn't safe to drink, forsake the ice, too. Many birds love bugs and some bugs love birders. *Insect repellent is generally a necessity* though some birders would dispute this, never having had much of a problem. Better to have it with you than to wish you had it.

SAFETY

There *is* safety in numbers. In most birding places in North America, the solo birder is perfectly safe. However, if you will be in an unfamiliar area, ask a local about personal safety. A rare bird of great interest to birders all across the continent was being sighted at the end of a deserted road. The rare bird alert cautioned birders about going there alone. It turned out that the area was a hangout for a motorcycle gang, members of which just might have hassled a lone birder.

CLOTHING

Clothing on all birding trips should be comfortable. Subdued colors are recommended. Comfortable shoes are a must for the active traveler-birder. Everywhere, wear closedtoe shoes rather than open sandals. Good walking shoes or boots will be welcome on tough terrain. It is often necessary to hike fairly long distances to see the "good" birds. Sometimes waterproof boots are indicated. An alternative is a pair of sneakers you don't mind getting wet. Extremes in temperature will be encountered in many places and "layering" is the general rule. Most birders carry a day pack in which to take a snack and to stuff the sweater as the day warms up.

BIRDER'S CHECKLIST

Experienced travelers know, as a result of experience, what things must be packed in order to make life tolerable. Just as a reminder, here's a checklist; the items you take will be dependent on whether the trip is for a few hours or a few months.

Binoculars (always first on the list)
Telescope and tripod if you think they'll be useful
Bird Guide(s) (but don't take the library)
Bird checklist
Other guidebooks depending upon interest (butterflies, mammals, wild-flowers, etc.)
Notebooks—a small one for the field, a larger one if you keep a log
Pencils, pens, and highlighter
Tape recorder and tapes
Kleenex and/or toilet paper; either may be nonexistent when you most need it
Hat
Sunscreen
Insect repellent
Water bottle
Flashlight with extra batteries
Alarm clock/watch
Wash cloth in a plastic bag; maybe an old towel
Extra plastic bags (for muddy boots and socks that didn't get dry, or to cover the camera exposed to road dust or ocean spray)
Poncho, the kind that folds up compactly, or other raingear
Scarf/bandanna
Spare eyeglasses, and prescription for same, if you can't read your bird guide without them

Survival kit (for the really well prepared): string, glue, rubber bands, spare shoe laces, needle and thread, scissors, small pencil sharpener, safety pins, scotch tape)

For picnickers: camp stove, lantern, extra propane, small skillet, coffee pot, grill, charcoal, lighter, pot holders, soap, plastic bags, thermos

Weather radio

Birders quickly learn about the "hardware" they need for happy birding. The "software," keen eyes and ears, an alert and inquiring mind, basic common sense, and a good sense of humor, is also very necessary for a lifetime of good birding.

IMPORTANT ADDRESSES FOR BIRDERS

American Birding Association
P.O. Box 6599
Colorado Springs, CO 80934

Hawk Migration Association of North America (HMANA)
P.O. Box 3482, Rivermont Station
Lynchburg, VA 24503

International Council for Bird Preservation
United States Section
801 Pennsylvania Avenue, SE
Washington, DC 20003

National Audubon Society
950 Third Avenue
New York, NY 10022

Nature Conservancy
Suite 800
1800 North Kent Street
Arlington, VA 22209

Ornithological Societies of North America
P.O. Box 21618
Columbus, OH 43221-0618

World Wildlife Fund
1255 23d Street NW
Washington, DC 20037

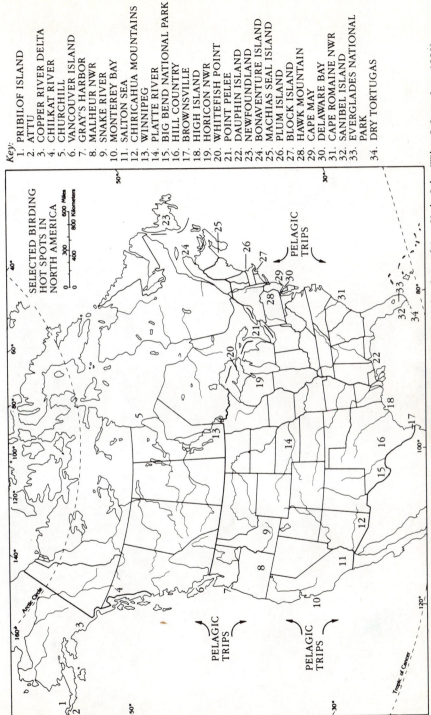

Key:
1. PRIBILOF ISLAND
2. ATTU
3. COPPER RIVER DELTA
4. CHILKAT RIVER
5. CHURCHILL
6. VANCOUVER ISLAND
7. GRAY'S HARBOR
8. MALHEUR NWR
9. SNAKE RIVER
10. MONTEREY BAY
11. SALTON SEA
12. CHIRICAHUA MOUNTAINS
13. WINNIPEG
14. PLATTE RIVER
15. BIG BEND NATIONAL PARK
16. HILL COUNTRY
17. BROWNSVILLE
18. HIGH ISLAND
19. HORICON NWR
20. WHITEFISH POINT
21. POINT PELEE
22. DAUPHIN ISLAND
23. NEWFOUNDLAND
24. BONAVENTURE ISLAND
25. MACHIAS SEAL ISLAND
26. PLUM ISLAND
27. BLOCK ISLAND
28. HAWK MOUNTAIN
29. CAPE MAY
30. DELAWARE BAY
31. CAPE ROMAINE NWR
32. SANIBEL ISLAND
33. EVERGLADES NATIONAL PARK
34. DRY TORTUGAS

SELECTED BIRDING HOT SPOTS IN NORTH AMERICA

0   100 300 500 Miles
0 200 400 600 800 Kilometers

PELAGIC TRIPS

Source: From *Study Guide to Accompany Geography: Regions and Concepts: 5th edition* (New York: John Wiley & Sons, Inc., 1988).

Bibliographic Summary

*P*roviding a formal bibliography would unnecessarily duplicate the list of books and periodicals included in Appendix 2. Many of these provided, at a minimum, some valuable nugget of information helpful in the preparation of this book. Much useful material was included on hundreds of national wildlife refuges, parks, and nature preserves. The Hammond Citation *World Atlas* (Maplewood, New Jersey, 1982) along with various maps of the world, of North America, and specific states were in constant use.

Primary bird checklists utilized for reference purposes (all listed in Appendix 2) included three major checklists: James Clements' *Birds of the World: A Checklist* (3d ed.); Howard and Moore's *A Complete Checklist of the Birds of the World*; and the American Birding Association's *A.B.A. Checklist* (3d ed.).

Other general reference works utilized included The *Audubon Society Encyclopedia of North American Birds,* Christopher Leahy's *The Bird Watcher's Companion,* and Ernest A. Choate's *The Dictionary of American Bird Names* (rev. ed.). An invaluable source of information was the National Geographic Society's *A Guide to Our Federal Lands.*

Although several North American field guides were referred to, the principal one in constant use during the preparation of this work was the National Geographic Society *Field Guide to the Birds of North America.* The several *Birder's Guides,* collectively known as the Lane guides, were used extensively, particularly for their calendar data. Peter Harrison's *Seabirds: An Identification Guide* was an essential reference book as was Hayman, Marchant, Prater's *Shorebirds.*

Reference books pertaining to individual states were not included in the book list, but Arnold Small's *The Birds of California* (New York: Winchester Press, 1974) was most useful in preparing sections relating to that state.

Reference has been made to a major article on migration appearing in *National Geographic* magazine, August 1979 (National Geographic Society, Washington, DC)

Much useful material was derived from the catalogs and newsletters of several birding tour operators, notably Field Guides, Inc.; Victor Emanuel Nature Tours (VENT), and Wings, Inc.

Glossary

Anatomical parts of birds are not included; every standard bird guide contains a labeled diagram of the standard bird.

ACCIDENTAL. Describes species that occurs in a particular place very rarely, or infrequently very far outside its normal range. See also RARITY.

AMERICAN ORNITHOLOGICAL UNION (AOU). The scientific organization that publishes the authoritative *A.O.U. Check-list of North American Birds,* on which most field guides are based. An AOU committee makes periodic decisions on combining or separating North American species (lumping and splitting) based on the latest scientific knowledge and makes official changes in the American names of birds. Publishes *The Auk,* a quarterly journal.

ATLAS. A census of breeding birds, occurring in given geographic areas, generally of a state. An atlas utilizes a system of mapping the area covered to determine numbers of individuals within a species in the area. Atlas areas are observed, generally by volunteers, and results are published.

AUDUBON SOCIETY. Originally identified as a "bird-watching" organization. Audubon Societies at the national, state, and local level today, focus their attention on wide-ranging environmental issues. The National Audubon Society publishes *Audubon* magazine and *American Birds.*

BIRD OF PREY. Generally refers to Osprey, hawks, eagles, and falcons, sometimes to vultures and owls.

CANOPY FEEDING. 1. A peculiar habit of some herons to hold their wings out to shade the water around them presumably to attract fish to the ensuing shade. The Reddish Egret can be distinguished at a great distance by this behavior. Other egrets, particularly the Louisiana Heron, occasionally exhibit this behavior. 2. The habit of vast numbers of birds in the tropics that feed only in the crown or canopy of trees, rarely descending to afford birders' good looks at them.

CENSUS. A periodic count of the numbers of birds within each species in a given geographic area. Periodic census counts, such as the annual Christmas Bird Count, provide time-lines for determining changes in bird populations.

CONSPECIFIC. Belonging to the same species. For example, formerly two separate species, the Mexican Duck is now considered conspecific with the Mallard.

DIHEDRAL. In birds, the angle at which the wing meets the body. A good reference point is the angle at which Turkey Vultures' wings are raised when soaring. The Black Vulture's dihedral is less pronounced; the Bald Eagle's is virtually flat.

DIMORPHISM. Occurrence of two distinct forms. Sexual dimorphism describes different plumages of the male and female of a species. Dimorphism may also refer to differences in characteristics such as size between male and female. See MORPH.

DISPLAY. Any activity engaged in by a bird designed to induce a desired behavior by another bird or mammal. Some shorebirds do a "broken wing" display to lure intruders away from a nest. Males often display their feathers or engage in specific behavior designed to attract the female. Males of some species construct elaborate display grounds for such activity. See LEK.

DUCK STAMPS. Annual stamps issued by the U.S. Fish and Wildlife Service, the purchase of which is required of all duck hunters. Proceeds of the stamps, known for their artistic excellence, finances the purchase of lands to be utilized as wildlife refuges, most of which are located along major flyways. Holders of duck stamps are entitled to free entry into refuges requiring an admission fee.

ECOLOGICAL NICHE. (See NICHE)

ENDANGERED SPECIES. Species identified in North America by the U. S. Fish and Wildlife Service as being "in danger of extinction throughout all or a significant proportion of its range." The California Condor is now extinct in the wild. Endangered species around the world are listed in the Red Data Book, published periodically by the International Council for Bird Preservation.

ENDEMIC. Restricted to, or found only in, a particular geographic area. Smith's Longspur is endemic to North America although the Lapland Longspur is Holarctic (Circumpolar).

EXOTIC. An introduced species flying free but not generally well adapted to the environment into which it has been introduced. These are often escapees.

FAUNAL. Relates to animal life of an area. Faunal regions—see Zoogeographic.

FLYWAY. The air-highway over prominent topographic features along which large numbers (but certainly not all) of birds migrate. The Atlantic and Pacific flyways follow the coast, the Mississippi flyway follows that river, and the Central flyway proceeds north from central Texas.

GOLDEN EAGLE PASS. Annually issued pass which allows entrance to all National Parks in the United States that require a fee. Golden Age Passports are issued free to citizens 62 years of age or older entitling them to lifetime park entrance. Both also entitle free entrance to national wildlife refuges.

LEK. A place where a number of male birds gather to display for the females. Grouse utilize leks.

LIFE ZONES. North American designation of areas of roughly 400 miles of north-south direction, or 2500 feet of elevation, within which a particular combination of flora and fauna may be found. Ranges from lowest and hottest to highest and coldest. Bird guides often provide maps showing life zones.

MELANISM. Unusual dark coloration of a bird's plumage. Also known as "dark phase," particularly common with hawks.

MIST NET. A finely woven net, set up on poles close to vegetation, into which birds fly. Birds become harmlessly caught in the net and can be removed for banding or other study purposes before being released.

MONOTYPIC. A taxonomic category which contains only one representative of the next lowest category. For example, a monotypic order would contain only one family; a monotypic family would contain only one genus. The Osprey is monotypic; it is the only species in the family Pandionidae.

MORPH. A phase or variation of color, size, or other characteristics that is different from the norm. Many hawks are recognized as dark phase or light phase. The Snow Goose has two color phases: white and blue. Red phase Eastern Screech-Owls are more commonly found in the South than are gray phase owls. Species are known as POLYMORPHIC if they have two or more phases (e.g., light, dark, and intermediate).

NICHE. The place or function an organism plays in the eco-system. Auks (family Alcidae) in northern oceans fill the ecological niche filled by Penguins (family Spheniscidae) in southern oceans.

NIDIFICATION. Relates to nests and nest building.

NOMINATE. In effect, the first officially documented species in a genus. The Latin name may repeat the name (e.g., the Eastern Kingbird, *Tyrannus tyrannus,* the nominate species of the genus Tyrannus). A good tour leader might urge you to become familiar with the nominate species in order to better identify other related species.

ORNITHOLOGY. Scientific study of bird life. A birder is not an ornithologist unless he or she has acquired the scientific background prescribed for that discipline. Amateurs routinely make considerable contributions to certain facets of ornithology to an extent not customary in other scientific disciplines.

PARASITIZE. Practice of some species of laying their eggs in another species nest. Host birds give preference to the "foreigner" thus neglecting their own brood. Cowbirds are notorious for this practice.

PASSERINE. Member of the order Passeriformes, the largest of the orders; includes half the bird species. In North America this is roughly translated as "perching bird." The order is derived from the Latin *passer* for sparrow. British birders define the term as meaning "sparrowlike."

PELAGIC. Relates to the ocean or open sea rather than to coastal waters. Pelagic birds spend their lives on the ocean, coming ashore only to breed and nest.

PHYLOGENIC. Roughly the evolutionary sequence of an order, family or species arising from a common ancestor. Virtually synonymous with *Taxonomic.* The sequence of listing of species in a bird guide is by the phylogenic or taxonomic order.

RAPTOR. A bird of prey (e.g., hawks, eagles, falcons, and sometimes owls).

RARITY. A bird uncommon in a particular area although it may be common elsewhere.

RIPARIAN. Relates to plants or animals living in areas influenced by the presence of rivers, streams, bogs, ponds, or lakes. In the dry western landscape, the presence of such wet areas often may be discerned by the presence of trees and bushes.

ROOKERY. Nesting colonies of birds, such as herons, often on islands.

SONOGRAM. Visual representation of what a bird's song is like. Shown in some bird guides.

TAXONOMY. The systematic ordering of plants and animals based on similarities and differences. Birds are sorted by orders, families, genera, and species.

TUBENOSE. An extra nostril, or salt gland, which enables sea birds to extract and excrete salt from saltwater, thus enabling them to live far from a source of fresh water. Refers to members of the order Procelariiformes: the albatrosses, shearwaters, petrels, and storm petrels.

ZOOGEOGRAPHIC. Relates to the distribution of flora and fauna of large geographic areas (see Chapter 2).

Index

ABA, *see* American Birding Association
Accentor, Siberian, 40
Alaska, 71–72, 76–83, 140–141. *See also*
    place names
Albatross:
    Black-footed, 71, 98, 120
    Laysan, 166
    Short-tailed, 80
Alberta, 91, 116
Alden Peter, 29, 48, 169, 209
American Birding Association, 11–13, 57
    A.B.A. Checklist, 20, 22
    annual convention, 86
    annual lists, 11
    definition of North America, 22
American Birds, 5, 30, 52, 59, 62, 70, 98,
    109, 113–114, 137
    pelagic trip, 112
American Ornithologists' Union, 22
    definition of North America, 22
Amherst and Wolf Islands, 140
Ani:
    Grooved-billed, 122
    Smooth-billed, 157
Año Nuevo State Reserve, 151
A.O.U., *see* American Ornithologists'
    Union
Appel, Captain Ted, 132
Arctic breeding areas, 82
Arctic mammals, 83, 85
Arizona, 55–57, 94–95, 136, 154–156.
    See also specific place names
Arizona Sonora Desert Museum, 225
Atlantic, 15
Attu Island, 40, 78–79
Audubon Canyon Ranch, 151
Audubon Corkscrew Swamp Sanctuary,
    96, 131
Auklet:
    Cassin's, 120
    Crested, 141
    Parakeet, 141
    Rhinoceros, 120, 163

    Whiskered, 80
Avocet, American, 91

Baffin Island, 82
Baja California, 137–138
Balch, Larry, 78
Bananaquit, 131
Baringer, Jill, 210
Basham, Benton, 12, 37, 78
Beach, Paul, 106, 139
Becard, Rose-throated, 95
Bellbird, Three-wattled, 61
Bentsen State Park, 53, 60, 122, 134, 153
Berkey, Gordon B., 30
Bermuda, 22
Big Morongo Wildlife Reserve, CA, 151
Binoculars, 209–210
Biotic communities, 23–28
Bird Observer tour group, 65
Bird population, by state, 147
Bird Watcher's Digest, 12
Birder's World, 12, 112, 155, 217
Birding, 12, 212
Bird-finding guides. *See also* Appendix 2
    Finley, 89, 91
    Lane, 5, 53, 93, 150, 152
    Pettingill, 55
Birding tour operators, 16–17. *See also*
    Appendix 2
Bittern, Least, 70
Blackbird:
    Tricolored, 136, 149
    Yellow-headed, 136
Blackpoll, 31, 104
Block Island, 113
Bluebird: Eastern and Mountain, 86
Bluenose Ferry, 88
Bluetail, Red-flanked, 79
Bluethroat, 80
Bobolink, 45, 94
Bonaventure Island, 87
Bond, James, 61, 168
Bonney, Jr., Richard E., 210

Booby:
  Brown, 65, 166
  Masked, 65, 98–99, 143
  Red-footed, 166
Boot Spring Canyon, 54
Bouton, Bill, 35, 58, 78, 95
Boyle, William J., Jr., 109
Brambling, 72
Brant, 29, 82
Bream, Bert, 215
British Columbia, 91, 116–117, 141, 163.
  See also specific place names
British Columbia Waterfowl Society,
  163
Brownsville, 37
  city dump, 39, 133
Buck, Mary, 169
Budgerigars, 158
Bufflehead, 162
Bulbul, Red-whiskered, 38, 157
Bunting:
  Blue, 123
  Indigo, 70
  Lark, 25, 91, 113
  Painted, 50, 131
  Rustic, 142
  Snow, 115, 140
  Varied, 54
  Yellow-breasted, 79
Burrows, Edward 128
Bushtit, Common, 26, 96, 141, 148,
  153, 155
Butterfly, Monarch, 123

California, 70, 118–121, 136–137,
  148–152. See also specific place
  names
Canada, 86–91. See also specific place
  names
Canadian Wildlife Service, 32
Canvasback, 59, 116, 162
Cape Ann, 139
Cape Cod, 94
Cape Hatteras, see Outer Banks
Cape May, New Jersey, 15, 16, 67,
  105–109
  Bird Observatory, 106–107
Caracara, Crested, 153
Caribbean Islands, 21–22
Carlsbad Caverns, 55
Carlton, Peter, 58
Carmel, CA, 150
Cascade Mountains, 117
Castellow Hammock Park, 131
Catbird, Gray, 108
Cave Creek Canyon, 55, 95
Central America, 22
Chacalaca, Plain, 134, 153
Chaffinch, Common, 72
Chat, Yellow-breasted, 93, 98
Checklists, bird, 5–6, 14–15. See also
  Appendix 2
Chesapeake Bay, 113–114
Chesterfield, Norm, 11
Cheyenne Bottoms State Wildlife
  Management Area, 164

Chickadee:
  Boreal, 86, 89–90, 112, 139
  Carolina, 49
  Mexican, 56
  Mountain, 90, 92
Chilkat River, 141
Chincoteague, 160
Chiricahua Mountains, 39, 55, 155
Christmas Bird Count, 8, 127–129
Christmas Bird Count, local, 128–129,
  135, 137, 139–141, 164
Chuck-will's Widow, 157
Chukar, 151
Churchill:
  birding in, 83–85
  polar bears in, 115–116
Clements, James F., 138
  world bird checklist, 22. See also
  Appendix 2
Colorado, 62, 92, 163–164. See also
  specific place names
Columbia River Basin, 71
Commanche National Grasslands, 62
Computerized Bird Lists, 213–214
Cooper, Mort, 157
Coot, American, 141
Copper River Delta, 71
Corella, Little, 135
Corkscrew Swamp Sanctuary, 96, 131
Cormorant:
  Brandt's, 121, 136, 151
  Double-crested, 76
  Great, 76, 139
  Pelagic, 121, 136, 151
  Red-faced, 81
Costa Rica, birds of, 61
Cowbird, Shiny, 37, 168
Crane:
  Sandhill, 46–47, 64, 70, 116, 122, 131,
    135, 151, 164
  Whooping, 46, 64, 92, 116, 132, 136
Crane Creek State Park, 69
Crossbill:
  Red, 25, 89–90
  White-winged, 89
Crow, Mexican, 39 52, 122, 133
Cuckoo:
  Black-billed, 68, 70, 89–90
  Common, 78
  Mangrove, 15, 65, 97, 155
  Oriental, 40, 79
  Yellow-billed, 15, 89–90
Curlew:
  Bristle-thighed, 28, 80, 146
  Eurasian, 40
  Little, 123
  Long-billed, 50, 107, 151
Cutler, David A., 109

Darling, J. N. "Ding," 96. See also
  National Wildlife Refuges
Dauphin Island, 49
Davidson, Mary, 79
Davis Mountains, 55
Delaware Bay, 67, 106, 113
Denver Field Ornithologists, 163

Dickcissel, 69
Dipper, American, 92
Dotterel, Eurasian, 78
Douglas, Marjorie Stoneman, 130
Dove:
 Collared, 38–158
 Inca, 153, 155
 Key West Quail-, 10, 155
 Ruddy Ground-, 133
 Spotted, 149
 White-winged, 59
 Zenaida, 97
Dovekie, 139
Dowitcher:
 Long-billed, 51,164
 Short-billed, 48, 51
Drennan, Susan Roney, 128
Dry Tortugas, 17, 64–65
Duck:
 Black-bellied Whistling-, 37,
  49
 Fulvous Whistling-, 49, 65, 131
 Harlequin, 121, 139, 151
 Masked, 121
 Mottled, 96
 Muscovy, 38
 Ruddy, 66
 Tufted, 78, 140
 Wood, 132, 162
Duck Stamps, 30, 97
Dunes State Park, 69
Dunlin, 22
Dunn, Pete, 108

Eagle:
 Bald, 15, 21–22, 48, 63, 94, 105,
  131–132, 141, 163–165
 Golden, 35, 63, 105
 White-tailed, 79
Edwards Plateau, 54–55
Egret:
 Cattle, 159
 Chinese, 40
 Great, 49
 Reddish, 48, 96, 155
 Snowy, 40
Ehrlich, Paul R., 32
Eider, 81–82
 Common, 85, 139
 King, 82
 Steller's, 141
Ellesmere Island, 20
Emanuel, Victor, Texas birding, 121
Evanson, Randall M., 70
Exotic birds, South Florida, 157–159

Falcon:
 Peregrine, 85, 105–111
 Prairie, 59–60, 152, 164
Falcon Dam, 135
Fall-outs, 52
Families of birds, See Appendix 1
Farallon Islands, 119, 151
Field Guides, 14, 16, 18, 52, 62. See also
 Appendix 1
 order of birds, 27, 31

Field Guides Inc., 103, 140. See also
 Appendix 2
Fieldfare, 73, 140
Finch:
 Cassin's, 91
 Rosy, 25, 61, 164
Fine, Rob and Shirley, 87
Flamingo, Greater, 131
Florida, 64–65, 96–97, 130–132, 156–157
Flycatcher:
 Acadian, 32
 Brown-crested, 151
 Buff-breasted, 95
 Dusky, 57, 92
 Dusky-capped, 95
 Fork-tailed, 99
 Gray, 57, 155
 Great-crested, 32
 Hammond's, 57, 155
 LaSagra's, 146, 159
 Least, 68
 Nuttings, 143
 Olive-sided, 59, 89, 94
 Red-breasted, 78
 Siberian, 40
 Sulphur-bellied, 95
 Vermillion, 54, 155
 Willow, 70
 Yellow-bellied, 32, 59, 89
Forster, Richard A., 113
Fort Jefferson, 64–65
Four Points tour group, 65, 93
Francis Marion State Forest, 162
Frank, Charles, 48
Frigatebird:
 Great, 166
 Magnificent, 113
Fulmar, Northern, 32, 81

Gadwall, 116
Gallagher, Tim, 211
Gallinule, Purple, 157
Gannet, Northern, 32, 65–66, 87, 109
Gargany, 99
Gaspe' Peninsula, 87
Gemmill, Daphne, 14
Glacier Bay, 83
Gnatcatcher:
 Black-capped, 57
 Black-tailed, 119, 136, 155
 Blue-gray, 51
Godwit:
 Bar-tailed, 78, 100
 Hudsonian, 45, 53, 84
 Marbled, 67, 86, 91
Goldeneye:
 Barrow's, 63, 139
 Common, 139
Golden Gate Audubon Society, 151
Goldfinch:
 Eurasian, 73
 Lawrence's, 136
 Lesser, 136, 155
Goose:
 Canada, 22, 46, 70, 142, 160, 163–165
 Emperor, 137, 141

Greater White-fronted, 46, 133
Greylag, 78
Ross', 115, 133, 135–137
Snow, 82, 114, 116, 133–135, 141
Goshawk, Northern, 93, 105, 115, 139
Grackle:
Boat-tailed, 152
Great-tailed, 152
Grassquit, Black-faced, 131
Great Lakes states, 59, 69, 94, 115, 165
Grebe:
Clark's, 62–63, 119, 136, 141
Eared, 63, 137
Horned, 63, 86
Red-necked, 59, 86–87
Western, 62–63, 119, 136, 141
Greenfinch, Oriental, 40
Greenland, definition of North America, 20
Greenshank, Common, 140
Griffith, Jim, 16
Grosbeak:
Black-headed, 96
Crimson-collared, 123, 133
Evening, 25, 92, 165
Pine, 25, 86, 91, 93, 115, 140, 165
Rose-breasted, 70, 90
Grouse:
Ruffed, 115
Sage, 26, 62, 91
Sharp-tailed, 46, 62, 115–116
Spruce, 83, 112, 115, 139
Guadaloupe Canyon, 57
Guan:
Black, 61
Crested, 139
Guillemot:
Black, 87, 139
Pigeon, 120, 151, 163
Gulf Coast, 45–52, 132–133
Gull:
Common Black-headed, 80, 139, 160
Franklin's, 116, 122
Glaucous, 32, 82, 134, 139
Glaucous-winged, 83, 150, 163
Heermann's, 119–120, 150
Herring, 87
Iceland, 139
Ivory, 81–82
Laughing, 119
Lesser Black-backed, 123, 132, 134
Little, 72, 123, 160
Mew, 83, 119–121, 150
Ring-billed, 116
Ross', 5, 70, 84
Sabines, 82, 120
Slaty-backed, 82, 142
Thayer's, 82, 119, 134, 150
Yellow-footed, 119, 151
Gyrfalcon, 59, 82, 115–116, 142

Hamel, Peter J., 57
Harrier, Northern, 60, 86, 105
Hartford Audubon Society, 84
Hawaiian Islands, 21–22, 165
Hawk:

Black-chested, 61
Broad-winged, 52, 59–60, 105–111
Common Black, 55, 95
Cooper's, 60, 111
Crane, 123, 143
Ferruginous, 60, 152
Gray, 95, 135
Harris, 153
Red-shouldered, 105–111, 152
Red-tailed, 60, 105–110, 118, 165
Roadside, 133
Rough-legged, 60, 70, 164–465
Sharp-shinned, 59–60, 105–107, 118
Short-tailed, 105
Swainson's, 52, 59–60, 86, 105, 152
White-tailed, 54
Zone-tailed, 95
Hawk Migration Association of North
America, 60, 106–111, 141
Hawk Mountain Sanctuary, 109
Hawk Ridge Nature Preserve, Duluth, 115
Hawk watching, 59–60, 104–111
Heron:
Green-backed, 136
Western Reef, 39
High Island, 52–53
Highway rest stops, 34–35
HMANA, see Hawk Migration Association
of North America
Hokanson, Harry and Ginny, 15
Holleyman Nature Reserve, 49
Honeycreepers, 166
Houston Audubon Society, nature
reserves, 52
Hummingbird:
Anna's, 136, 141, 148
Berylline, 95
Black-chinned, 95
Blue-throated, 95
Broad-billed, 95
Broad-tailed, 55, 95
Buff-bellied, 122, 153
Caliope, 91, 93, 95
Costa's, 149
Lucifer, 54–96
Magnificent, 95
Ruby-throated, 32, 50, 152
Rufous, 95, 113
Violet-crowned, 95
White-eared, 95
Xantus, 143
Huron National Forest, 58

Ibis:
Glossy, 49
White, 161
White-faced, 49, 153
Idaho, 63–64, 92. See also specific place
names
Indiana, 69–70

Jacana, Northern, 121
Jackdaw, Eurasian, 100
Jaeger:
Long-tailed 83

Parasitic, 71, 84, 120
Pomarine, 32, 71
Jamaica, birds of, 61
Jamaica Bay Wildlife Refuge, 40
Jasper-Pulaski State Fish and Wildlife
    Area, 70
Jay:
    Gray, 89–90, 117, 140
    Green, 122, 153
    Mexican, 155
    Pinyon, 26, 164
    Scrub, 26, 136
    Stellar's, 117
Jekyl Island, 114
Junco:
    Dark-eyed, 164
    Yellow-eyed, 57, 155

Kansas, 68, 164
Kaufman, Kenn, 15, 155
Kelley, Robert, 11
Kestrel:
    American, 107–108
    Eurasian, 40
Key Biscayne, 65
Key West, 96
Kingbird:
    Couch's, 122
    Eastern, 45, 50
    Gray, 65, 97, 156
    Thick-billed, 57, 95
Kingery, Hugh E., 62
Kingfisher:
    Green, 35, 136, 153
    Ringed, 39, 122, 135, 153
Kinglet, Ruby-crowned, 25
Kisatchie National Forest, 49
Kiskadee, Great, 52, 122, 153
Kite:
    Black-shouldered, 136, 153
    Hook-billed, 123, 133
    Mississippi, 52–53
    Snail Kite, 47, 96
    Swallow-tailed, 65, 131, 155
Kittiwake:
    Black-legged, 81, 87, 160
    Red-legged, 81
Klamath Basin-Tule Lake, 163
Knot, Red, 67
Kodiak, 140–141
Komito, Sanford, 12, 15

Lake Okeechobee, flooded fields, 97, 103
Lane, James, 5. See also Bird-finding
    guides; Appendix 2
Lapwing, Northern, 73
Lasley, Greg W., 52–53
Leks, 54, 62
Leo, John, 12, 112
Limpkin, 157
Lipsky, Mike, 2
Listing, 13–15
    checklists 5–6. See also Appendix 2
    Big Year, 15
Living Bird Quarterly, 12, 210

Long Island, 67
Long Point, 59
Longspur:
    Chestnut-collared, 25, 34, 63, 136
    Lapland, 46, 83
    McCown's, 25, 63
    Smith's, 46
Loon:
    Common, 114
    Pacific, 59, 71, 85
    Red-necked, 85
    Yellow-billed, 59, 71, 82, 141, 162
Los Angeles Audubon Society, 150
Louisiana coast, 48–50
Louisiana State Museum, bird skin
    collection, 48

Machias Seal Island, 87
Madera Canyon, 57, 95
Magpie:
    Black-billed, 172
    Yellow-billed, 137, 172
Maine, See specific place names
Mammals:
    Arctic, 83, 85
    Yellowstone, 92
Manitoba, 34, 70, 86. See also Churchill;
    Winnipeg
Maritime Provinces, 87–88, 112. See also
    specific place names
Martha's Vineyard, 113
Massachusetts, 73, 94, 139. See also
    specific place names
Maxham, Maida, 137, 217
McAllen, 133
McCormick, Myra, 90
McKaskie, Guy, 137
Meadowlark, Eastern/Western, 69, 93, 152
Merganser:
    Hooded, 66
    Red-breasted, 66
Meriwether, William M., 212
Merlin, 59, 105
Mexican vagrants, 121
Mexico, birding in, 122, 138
Michigan, 58
Mickelson, Belle and Pete, 71
Midwest, 68–69, 114
Migrant traps, 49, 112. See also Salton Sea
Migration, 6
    flyways, 28–33, 47, 68–70, 103–104,
        114–115, 117, 164
    effects of weather, 30–31, 58, 103, 109
    post-breeding reverse, 119
Minnesota, 86, 115, 165
Mockingbird, Bahama, 73, 97, 159
Monhegan Island, 68, 112
Montana, 93
Monterey Bay Aquarium, 149
Monterey Peninsula, 39, 119
Moriarty, Dan, 166
Morse, Robert J., 56
Murrelet:
    Craveri's, 120, 123
    Kittlitz's, 83

Marbled, 83, 120
Xantus, 71, 150
Murre:
    Common, 81, 87, 141, 162
    Thick-billed, 81

Nantucket Island, 39
NARBA, *see* North American Rare
    Bird Alert
Nash, Dolly, 108
National Audubon Society, publisher
    of *American Birds*, 5
    Sandhill Crane viewing area, 47
    Florida Keys office, 148
National Parks:
    Acadia, 94
    Big Bend, 54, 97
    Everglades, 37, 65
    Channel Islands, 150
    Grand Teton, 92
    Haleakala, 166
    Hawaiian Volcano, 166
    Mammoth Cave, 98
    Riding Mountain, 86
    Rocky Mountain, 92
    Yellowstonea, 92
    Wood Buffalo, 47, 132
National Wildlife, 3
National Wildlife Refuges:
    Anahuac, 51
    Aransas, 48, 132
    Attwater Prairie Chicken, 53
    Audubon, 116
    Bear Lake, 64
    Bombay Hook, 114
    Bosque del Apache, 92, 116, 135
    Brigantine, 108, 160
    Cape Romaine, 66, 162
    Delta, 132
    Erie, 59, 94
    Flint Hills, 68
    Grays Lake, 92
    Great Dismal Swamp, 67
    Great Meadows, 113
    Great Swamp, 67
    Horicon, 70, 165
    J. N. "Ding" Darling, 96
    Lacaissine, 49
    Lacreek, 93
    Lake Alice, 116
    Lewis and Clark, 142
    Loxahatchee, 65, 131
    Malheur, 91, 117
    Minidonka, 64
    Minnesota River Valley, 165
    Monomoy, 160
    Moosehorn, 67, 94
    Muleshoe, 135
    Parker River, 67, 139, 159
    Pea Island, 160
    Piedmont, 162
    Quivira, 164
    Red Rock Lakes, 92
    Rice Lake, 115
    Sabine, 49
    St. Marks, 49, 131

Santa Ana, 122, 134
Seney, 194
Souris 93
Swan Lake, 69
Tishomingo, 135
Turnbull, 118
Union Slough, 70
Upper Souris, 116
Washita, 122
Waubay, 68
Nature Conservancy:
    Mile Hi Sanctuary, 95, 155. *See also*
        Appendix 2
    Oregon bird walk, 91
    Sandhill Crane viewing area, 47
Nature Travel Service, 16
Nebraska, *See* Platte River
Neville, Bruce, 105, 155, 159
New Brunswick, 72
Newfoundland, 73, 87, 140
New Jersey, *See* Hawk watching
New Mexico, 55, 135
New York City, Central Park, 67
New York Times, bird reports, 28, 30
Niagra Falls, gull migration, 103
Nighthawk, Antillean, 65, 97, 156
Nightjar, Buff-collared, 55, 57
Noddy:
    Black, 65
    Brown, 65
Nome, 82
North America, geography of, 20–23
North American Rare Bird Alert, 13, 36,
    38, 40, 73, 79, 99, 142, 153
North Carolina, 98, 160. *See also* specific
    place names
North Dakota, 47, 93, 116
Northwest Territories, Banks and Bylot
    Islands, 82
Nova Scotia, 72–73
Nutcracker, Clark's, 28, 59, 90–91, 117
Nuthatch:
    Brown-headed, 49, 53, 66
    Pygmy, 56, 92
    Red-breasted, 91

Odear, Bob, 79 113
O'Neil, John, *Birds of Peru*, 48
Oregon, 117, 142
Oriole:
    Altamira, 52, 122, 134, 153
    Audubon's, 122
    Black-vented, 54
    Hooded, 15
    Northern, 32, 69, 165
    Orchard, 49, 53, 93
    Spot-breasted, 97, 155
    Streak-backed, 136
Osprey, 33, 60, 94, 105
Outer Banks, 72–73, 114, 160
Ovenbird, 98
Owl:
    Barn, 155
    Barred, 96
    Boreal, 59, 64, 100, 139, 163
    Burrowing, 91, 93, 149, 155

Eastern Screech, 140, 152
Elf, 26, 54, 95, 155
Ferruginous Pygmy, 100, 135
Flammulated, 95, 100
Great Gray, 4, 64, 85–86, 115, 140
Great Horned, 140, 155
Long-eared, 64
Northern Hawk, 85, 91, 100, 115,
    139–140
Northern Pygmy, 155
Northern Saw-whet, 64, 100, 115, 123,
    140–141
Short-eared, 64, 140
Snowy, 15, 81, 115–116, 137, 140
Spotted, 95, 151, 155
Western Screech, 54, 64, 95, 152, 155
Whiskered, 95, 155
Oystercatcher:
    American, 67
    Black, 119, 148

Palo Alto Baylands Reserve, 151
Parakeet:
    Canary-winged, 38, 157
    Green, 122, 133
    Monk, 158
Parker, Ted, 48
Parrot, Red-crowned, 122, 133, 155
Partridge, Gray, 64, 86, 116
Parula:
    Northern, 98
    Tropical, 38, 123, 134, 143
Patagonia, 57, 95
Pauraque, 122, 153
Pawnee National Grasslands, 62
Paxton, Robert O., 109
Pea Island, 114, 160–161
Pelagic birding:
    East coast, 66, 98, 161
    Maritime Provinces, 87, 112
    West coast, 71, 97–98, 119–121
Pelagic birds, 32–33
Pelican:
    American White, 35, 68, 93, 116–117,
        130
.   Brown, 67, 150, 162
Perry, Steve, 15, 143
Peterson, Roger Tory, 67, 84
    favorite birding areas, 117, 163, 165
Petrel:
    Black-capped, 98, 121, 123
    Cook's, 121
    Mottled, 121
    Solander's, 72
Pewee:
    Eastern/Western Wood-, 152
    Greater, 57
Phainopepla, 95, 155
Phalarope:
    Red, 66, 120
    Red-necked, 32, 120
    Wilson's, 62, 68, 91, 164
Philadelphia, 113
Phoebe:
    Black, 95, 148
    Eastern, 108

Say's, 113
Pigeon:
    Band-tailed, 96, 148–149, 155
    Red-billed, 52
    White-crowned, 57
    White-winged, 155
Pintail, Northern, 137
Pipit:
    Meadow, 22
    Pechora, 79
    Red-throated, 117
    Sprague's, 15, 54,59, 86, 91
    Water, 111, 119
Platte River, cranes at 46–47
Plover:
    Common Ringed, 82
    Greater Golden-, 73
    Lesser Golden-, 8, 29, 82, 84, 117
    Little Ringed, 78
    Mongolian, 79, 82
    Mountain, 136–137
Plum Island, see National Wildlife
    Refuge, Parker
Pochard, Common, 79
Point Pelee, 4, 16, 57–59, 110–111
Point Reyes National Seashore, 119, 150–151
Portal, 55
Powell, David J., 59
Powell, Peggy, 79
Prairie Chicken:
    Greater, 45–47, 53, 62
    Ralph E. Yeater Sanctuary, 69
    Taberville Refuge, 70
    Lesser, 62
Pratincole, Oriental, 40
Presque Isle State Park, 59
Pribilof Islands, 81
Ptarmigan:
    Rock, 82–83
    White-tailed, 90, 164
    Willow, 82, 85
Pterodroma 121
Puerto Rico birds, 61
Puffin:
    Atlantic, 87–88
    Horned, 81
    Tufted, 81, 120, 163
Pyrrhuloxia, 54, 155

Quail:
    California, 148
    Gambel's, 95, 153, 155
    Montezuma, 55, 136, 155
    Mountain, 148
    Scaled, 35, 153, 155
Quebec, 87, 89
Queen Elizabeth Islands, 82
Quetzal, Resplendant, 61

Rail:
    Black, 51, 94, 99, 109
    Clapper, 48
    Yellow, 51, 94, 99, 122
Rain forest, destruction of, 31–32
Rainwater Basin Wetland Management
    District, crane viewing, 47

Ramsey Canyon, 57
Rare Bird Alerts, 214–25
Rarities, *see* North American Rare
    Bird Alert
  finding, 10, 37–40
  reported, 72–73, 99–100, 122–123,
    142, 143,
  Texas, 154
Raven, Chihuahuan, 134, 155
Razorbill, 87, 139
Redhead, 70, 116
Redman, Nigel, 7
Redpoll:
  Common, 165
  Hoary, 139
Redshank, Spotted, 73
Redstart:
  American, 98
  Painted, 55–56, 155
Redwing, 140
Reed's Beach, 67
Reeve, *see* Ruff
Reifel, George C., Migratory Bird
    Sanctuary, 117, 163
Rhode Island Audubon Society, 113
Ribble, John and Barbara, 84
Rich, Terry, article on sparrows, 93
Rice Conservation Area, 114
Rio Grande River valley, 39, 53
River of Grass, 130
Roadrunner:
  Greater, 26, 52, 135, 153
  Lesser, 109
Robin:
  Clay-colored 133
  Rufous-backed, 54, 136
  Siberian Blue 40
Rocky Mountains:
  American, 91
  Canadian, 90
Rubythroat, Siberian, 40
Ruff, 38, 74, 78, 99, 117, 119, 159
Rustler Park, AZ 56
Rydell, Bill, 13

St. Lawrence Island, 81
St. Louis, 29, 142
Salton Sea, 71, 119, 137, 151
San Diego, 119, 150
San Francisco, 118, 150–151
Sanderling, 67, 97
Sandpiper:
  Baird's, 82, 86, 108, 164
  Buff-breasted, 82, 114, 117
  Curlew, 73, 82, 99, 109
  Green, 78
  Pectoral, 51, 82, 114
  Purple, 15, 32, 116, 139
  Rock, 81
  Semipalmated, 28, 67
  Sharp-tailed, 117, 119
  Spoonbill, 78
  Stilt, 84, 117, 164
  Terek, 78, 100

Upland, 45, 54, 70, 91, 93
  White-rumped, 51, 108, 164
  Wood, 78
Sanibel Island, 65, 96, 156
Santee State Park, 162
Sapsucker, Williamson's, 57, 91–92, 117
Savannah Coastal Refuges, 66
Scaup, Greater, 84
Scientific Names, *see* Appendix 1
Scopes, spotting, 210
Scops-Owl, Oriental, 40
Scoter:
  Black, 66, 151, 162
  Surf, 66, 151
  White-winged, 70
Seal, Northern Fur, 81
Seattle-Victoria area, 235
Seedeater, White-collared, 123
Seven Hundred Club, 9
Sexton, Chuck, 52
Shearwater:
  Audubon's, 65, 98
  Black-vented, 120
  Buller's, 98, 120
  Cory's, 98, 112
  Greater, 98, 104, 112
  Manx, 112
  Newell's, 166
  Pink-footed, 120
  Short-tailed, 123
  Sooty, 66, 102, 120
  Streaked, 39
  Wedge-tailed, 120, 166
Shearwater, Debi, 119–120
Sheyenne National Grasslands, 47
Shoveler, Northern, 137
Shrike, Northern, 115, 139
Sierra, 32
Siskin, Eurasian, 142
Skagit River delta, 141
Skimmer, Black, 67, 77, 97, 155
Skua:
  Great, 98
  South Polar, 28, 102, 120
Skylark, Eurasian, 162
Small, Arnold, 16, 118, 148
Snake River Birds of Prey Area, 63
Snetsinger, Phoebe, 142, 158
Solitaire, Townsend's, 59, 117, 136
Sonoita Creek Sanctuary, 95
South Dakota, 68, 93
Southern Mississippi, University of, 49
Sparrow:
  Bachman's, 49, 53
  Baird's, 91, 93
  Black-chinned, 26, 151
  Black-throated, 136, 151, 155
  Botteri's, 57, 95
  Brewer's, 69
  Cassin's, 54, 155
  Clay-colored, 52
  Eurasian Tree, 164
  Field, 69
  Five-striped, 57, 76, 100

Fox, 26
Golden-crowned, 83
Grasshopper, 54
Harris', 85, 93
Henslow's, 58–59, 113
Le Conte's, 54, 59, 113
Lincoln's, 52
Olive, 153
Rufous-crowned, 54, 155
Rufous-winged, 57, 95, 155
Sage, 26, 117, 136, 148
Sharp-tailed, 93
Swamp, 93
White-crowned, 83, 93, 111
White-throated, 111
Sparrows, North Dakota, 93
Special Odysseys, 82. *See also* Appendix 2
Species, world-wide, 32–33
Spoonbill, Roseate, 49, 65, 96, 131, 155
Starling, European, 141
Starthroat, Plain-capped, 95
States, bird species in, 147
Stenberg, Kate, 155
Stilt, Black-necked, 35, 49
Stint:
Rufous-necked, 40, 82, 99
Temminck's, 80, 117
Stork, Wood, 67, 96, 119, 131, 162
Storm-Petrel:
Ashy, 120
Band-rumped, 161
Black, 120
British, 20
Fork-tailed, 120
Leach's, 32, 88–89, 102, 162
Least, 123
Wedge-rumped, 40
White-faced, 112
Wilson's, 66, 88–89, 120, 163
Stuller, Stu, 15
Surfbird, 29, 73, 117, 119
Swallow:
Bahama, 65, 159
Cave, 555, 65
Rough-winged, 32, 111
Tree, 89, 108
Swan:
Mute, 108
Trumpeter, 63, 84, 91–93, 117, 141
Tundra, 22, 32, 59, 63, 69, 114,
141–142, 160–161, 163
Swift:
Black, 93, 151
Vaux's, 32, 71, 93
White-collared, 143
White-throated, 91, 137, 148, 155
Sycamore Canyon, 76

Tanager:
Flame–colored, 39, 56
Hepatic, 96
Scarlet, 50, 52
Summer, 50
Western, 71

Tattler, Wandering, 29, 117, 119, 150
Teal:
Cinnamon, 116
Eurasian Green-winged, 32, 72
Falcated Teal, 143
Green-winged, 32, 137
Teale, Edwin Way, 7, 181
Tennessee, 164
Tern:
Aleutian, 83
Arctic, 28, 83, 104, 120
Black, 70, 86, 93
Bridled, 65, 98–100
Elegant, 150
Forster's, 70, 86, 109
Least, 149, 97, 155
Roseate, 96
Royal 109
Sooty, 15, 65
White-Winged, 99
Texas, 51–54, 95–96, 121–123, 132–135,
152–153. *See also* specific place
names
Thick-knee, Double-striped, 61
Thompson:
John, 56, 114
Lydia, 56
Thrasher:
Bendire's, 57, 136, 151
California, 26, 148
Crissal, 16, 54, 136, 151, 155
Curve-billed, 136–137, 155
Le Conte's, 26, 136, 149
Sage, 26, 59, 117, 136–137, 155
Thrush:
Aztec, 54
Gray-cheeked, 111
Hermit, 53
Varied, 83, 90, 93
Wood, 50–51, 53
Time, article on birding, 9
Tingley, Stuart, 140
Tit, Siberian, 40
Titmouse:
Bridled, 26, 95, 155
Plain, 26
Tufted, 35
Toops, Judith A., 122
Towhee:
Abert's, 57, 136, 155
Brown, 26, 153, 155–156
Green-tailed, 26, 155
Travel & Leisure, 11
Trogon, Elegant, 39, 55–56, 95
Tropical Audubon Society, 157
Tropicbird:
Red-billed, 98–99, 113, 120
Red-tailed, 40, 166
White-tailed, 65, 166
Turkey, Wild, 153, 155
Turnstone:
Black, 29, 119, 136
Ruddy, 67, 97
Tyrannulet, Northern Beardless, 57, 95

Vagrants, 32, 38
Van der Geld, Ann, 141
Vancouver Island, 71–72
VENT, *see* Victor Emanuel Nature Tours
Verdin, 153
Victor Emanuel Nature Tours, 84, 117.
    See also Appendix 2
Vireo:
  Black-capped, 96
  Black-whiskered, 15, 65, 97, 155
  Gray, 54, 96
  Hutton's, 149, 155
  Philadelphia, 32
  Red-eyed, 31–32, 52
  Warbling, 71, 93
  White-eyed, 31
  Yellow-green, 99
Vulture, Turkey, 32, 55, 111

Wagtail:
  Black-backed, 100
  White, 82
  Yellow, 82
Warbler:
  Arctic, 28, 82
  Audubon's, 69
  Bay-breasted, 111
  Blackburnian, 52, 58
  Black-and-white, 111
  Black-throated Gray, 26, 151
  Black-throated Green, 52
  Blue-winged, 52, 90, 98
  Canada, 111
  Cape May, 111
  Cerulean, 52, 90, 98
  Chestnut-sided, 90, 104
  Colima, 54, 96
  Connecticut, 31, 58, 65, 86
  Fan-tailed, 100
  Golden-cheeked, 95
  Golden-crowned, 123, 133
  Golden-winged, 52, 89–90
  Grace's, 52
  Hooded, 32, 49, 53, 70, 90, 98
  Kentucky, 98
  Kirtland's, 58, 70
  Lucy's, 57
  MacGillivray's, 57, 71, 117
  Magnolia, 111
  Mourning, 58, 86, 90
  Myrtle, 69
  Nashville, 52, 58, 71
  Olive, 155
  Orange-crowned, 32
  Palm, 32
  Pine, 31
  Prothonotary, 52, 67, 98, 165
  Red-faced, 56, 95
  Rufous-capped, 54
  Swainson's, 52–53, 67
  Tennessee, 31, 52, 111
  Townsend's, 71, 90, 93, 117, 119
  Virginia's, 57

  Wilson's, 68, 71, 111
  Worm-eating, 49, 52, 67, 98
  Yellow-rumped, 69, 109, 152
  Yellow-throated, 98
  Yellow, 70, 98
Warren, Andy and Joan, 84
Waterfowl, decline in numbers, 30
Waterthrush:
  Louisiana, 98
  Northern, 21
Waxwing, Bohemian, 83, 90, 115, 137
Weir, Ron D., 31
West Indies, 21
Wheatear, Northern, 73, 80, 82, 113
Whimbrel, 22, 48
White, Gilbert, 6
Whitefish Point Bird Observatory, 59–60
Whooping Crane Maintenance Trust, 47
Wigeon, Eurasian, 78, 100, 115, 123
Wilcox, Helena, 165
Wild Bird, 12, 211
Wildlife refuges, United States and
    Canada, 30
Willet, 48
Williams, Frances, 30
WINGS, Inc., 88
Winnipeg, birding areas, 34, 86
Wisconsin, 70, 165
Wood, Tom and Mary, 58, 90
Woodcock, American, 67, 94
Woodpecker:
  Black-backed, 25, 86, 89, 137
  Downy, 24
  Gila, 26
  Ivory-billed Woodpecker, 38
  Ladder-backed, 35, 152
  Lewis', 91
  Northern Three-toed, 26, 86, 90–91, 93
  Nuttall's, 26, 148–149
  Pileated, 35, 38
  Red-bellied, 152
  Red-cockaded, 49, 53, 162
  Strickland's, 26, 136, 155
  White-headed, 24
Woodstar, Bahama, 159
Wren:
  Bewick's, 95, 136
  Cactus, 26, 35, 131, 153
  Canyon, 137, 153
  Marsh, 93, 136
  Rock, 59, 136, 149, 153
  Winter, 86
Wrentit, 25, 148
Wyoming, 92–93

Yellowlegs, Greater/Lesser, 51
Yellowthroat:
  Gray-crowned, 123, 134, 143
  Common, 50

Zoogeographic regions, 21